Math Drills

by
Bob Bernstein

illustrated by Bron Smith

Cover by Bron Smith

Copyright © Good Apple, 1992

Good Apple
1204 Buchanan St., Box 299
Carthage, IL 62321-0299

S I M O N & S C H U S T E R *A Paramount Communications Company*

Copyright © Good Apple, 1992

ISBN No. 0-86653-660-4

Printing No. 98765432

Good Apple
1204 Buchanan St., Box 299
Carthage, IL 62321-0299

Table of Contents

S Yoder

GA1392

Introduction

The drill activities presented in this book are specifically written to enhance both the teaching and development of computational skills while at the same time placing a strong emphasis on the application of thinking skills. It has been long thought of in the educational community that basic number facts can be learned and understood by students when they are not presented in isolation but rather when these same number facts are presented in a positive and relevant setting.

The basic goals of *Math Drillsters* are

1. to help students gain a strong, positive foothold in their early understanding and application of basic math computational skill proficiency

2. to encourage all students to enjoy mathematics by having them participate in drill activities that are challenging and creative

3. to improve the students' knowledge and application of math computational skills to the point where the students feel confident and begin to enjoy the world of mathematics

GA1392

4, 3, 2, 1

The 4, 3, 2, 1 is a practice drill for computational skill development. On every page, you will find forty problems. Included is a page for addition, subtraction, multiplication, division and a final set that includes a page with all four operations. The same problems can be found on the **4, 3, 2** and **1** pages. The difference in each page is the amount of time that is allotted to successfully complete the page. The student will have four minutes to complete the **4**, three minutes for the **3**, two minutes for the **2** and one minute for the **1**. Some type of stopwatch will be most helpful during the drill. The pages (4, 3, 2, 1) should be copied for each student.

Participants should begin with the forty problems (perhaps addition) that can be found on the large numeral **4**. Inform the students that this will be a timed drill. According to the stopwatch, they will be told when to begin and when to stop. Each student will have four minutes in which he will try to correctly solve every problem. At the completion of the four-minute period, students should exchange their papers with their neighbors. At this time the teacher will read the answers to the problems.

Example:
>   7 + 5 = 12,   9 + 1 = 10, etc.

4, 3, 2, 1

GA1392

Ask the children to record a *C* on each correct answer. When the teacher has completed the review, the students should total all correct answers in the space provided and also the amount of time that was allotted.

Practice the four-minute drill as many times as deemed necessary before going on to the three-minute drill. This same format should progress through to the two and one-minute drills.

The numerals with blank spaces are provided for a teacher's perception of the particular needs of his/her classes.

2

GA1392

Name _____
Correct Answers _____
Time _____

7 + 5 ☐	9 + 1 ☐	3 + 8 ☐
9 + 9 ☐	6 + 7 ☐	5 + 9 ☐
4 + 8 ☐	11 + 9 ☐	10 + 2 ☐
7 + 7 ☐	8 + 5 ☐	4 + 10 ☐
7 + 8 ☐	6 + 12 ☐	5 + 8 ☐
5 + 10 ☐	6 + 6 ☐	6 + 5 ☐

10
+ 0
☐

9 + 10 ☐	5 + 0 ☐	4 + 9 ☐
4 + 7 ☐	7 + 10 ☐	3 + 9 ☐
8 + 9 ☐	3 + 10 ☐	8 + 8 ☐
6 + 10 ☐	8 + 0 ☐	8 + 10 ☐
8 + 2 ☐	7 + 9 ☐	1 + 10 ☐
7 + 4 ☐	6 + 3 ☐	4 + 4 ☐
10 + 10 ☐	11 + 0 ☐	12 + 2 ☐

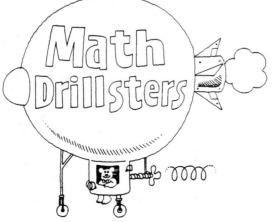

3

Name _____

Correct Answers _____

Time _____

12 − 4 ☐	10 − 2 ☐	11 − 9 ☐
9 − 9 ☐	12 − 11 ☐	10 − 6 ☐
11 − 0 ☐	10 − 7 ☐	12 − 10 ☐
12 − 9 ☐	9 − 7 ☐	11 − 1 ☐
9 − 2 ☐	12 − 2 ☐	10 − 3 ☐
12 − 0 ☐	11 − 10 ☐	10 − 9 ☐

11
− 11
☐

11 − 6 ☐	9 − 1 ☐	10 − 10 ☐
10 − 4 ☐	11 − 3 ☐	9 − 8 ☐
11 − 2 ☐	10 − 1 ☐	9 − 4 ☐
9 − 0 ☐	11 − 8 ☐	8 − 8 ☐
8 − 5 ☐	9 − 6 ☐	8 − 1 ☐
10 − 5 ☐	8 − 7 ☐	8 − 3 ☐
12 − 7 ☐	11 − 5 ☐	12 − 5 ☐

4

GA1392

Name _____
Correct Answers _____
Time _____

9 × 5 □	3 × 8 □	6 × 6 □
5 × 5 □	9 × 4 □	8 × 0 □
9 × 0 □	7 × 6 □	8 × 7 □
8 × 1 □	9 × 9 □	7 × 3 □
7 × 7 □	9 × 1 □	8 × 2 □
9 × 7 □	4 × 9 □	8 × 9 □

4
× 4
□

7 × 5 □	9 × 8 □	8 × 8 □
9 × 2 □	6 × 9 □	7 × 0 □
7 × 4 □	8 × 5 □	7 × 9 □
9 × 3 □	7 × 1 □	6 × 8 □
8 × 6 □	9 × 6 □	5 × 7 □
7 × 2 □	8 × 4 □	6 × 7 □
8 × 3 □	7 × 8 □	6 × 6 □

5

36-40 Super
31-35 Great
27-30 Good

Name _____
Correct Answers _____
Time _____

□
3⟌12

□
5⟌15

□
4⟌12

□
6⟌30

□
9⟌27

□
5⟌20

□
2⟌6

□
10⟌20

□
3⟌15

□
6⟌24

□
10⟌30

□
1⟌13

□
6⟌18

□
2⟌14

□
5⟌10

□
3⟌30

□
2⟌20

□
1⟌15

□
2⟌16

□
3⟌18

□
1⟌10

□
3⟌27

□
5⟌30

□
7⟌28

□
7⟌14

□
9⟌18

□
3⟌21

□
4⟌20

□
8⟌24

□
4⟌28

□
2⟌30

□
2⟌4

□
2⟌18

□
1⟌11

□
5⟌25

□
2⟌24

□
1⟌5

□
7⟌21

□
3⟌9

□
2⟌10

GA1392

Name _____
Correct Answers _____
Time _____

□ 7)21̄	9 × 4 □	8 + 9 □
12 + 0 □	12 − 8 □	10 − 5 □
6 × 6 □	□ 3)15̄	7 × 9 □
□ 2)14̄	□ 4)36̄	12 − 0 □
9 × 9 □	□ 5)15̄	12 − 4 □
□ 7)49̄	3 + 12 □	□ 8)16̄

8
× 10
□

□ 4)4̄	7 × 7 □	10 + 6 □
11 − 5 □	□ 3)18̄	9 × 8 □
10 × 3 □	2 × 10 □	11 − 11 □
12 + 0 □	6 × 8 □	4 + 9 □
□ 9)18̄	9 × 7 □	□ 9)63̄
11 − 1 □	□ 24)24̄	3 × 8 □
10 × 10 □	4 × 6 □	□ 7)28̄

80!

7

GA1392

36-40 Super
31-35 Great
27-30 Good

Name _____
Correct Answers _____
Time _____

This page is for you to create your own daily drill.

GA1392

Name _____
Correct Answers _____
Time _____

7	9	3	9	6	5	4
+ 5	+ 1	+ 8	+ 9	+ 7	+ 9	+ 8
□	□	□	□	□	□	□

11	10	7	8	4	7
+ 9	+ 2	+ 7	+ 5	+ 10	+ 8
□	□	□	□	□	□

6	5	5	6
+ 12	+ 8	+ 10	+ 6
□	□	□	□

6	10	9	5	4
+ 5	+ 0	+ 10	+ 0	+ 9
□	□	□	□	□

4	7	3
+ 7	+ 10	+ 9
□	□	□

8	3	8	6	8	8	8
+ 9	+ 10	+ 8	+ 10	+ 0	+ 10	+ 2
□	□	□	□	□	□	□

7	1	7	6	4	10	11	12
+ 9	+ 10	+ 4	+ 3	+ 4	+ 10	+ 0	+ 2
□	□	□	□	□	□	□	□

GA1392

Name _____
Correct Answers _____
Time _____

12	10	11	9	12	10	11
− 4	− 2	− 9	− 9	− 11	− 6	− 0
□	□	□	□	□	□	□

10	12	12	9	11	9
− 7	− 10	− 9	− 7	− 1	− 2
□	□	□	□	□	□

12	10	12	11
− 2	− 3	− 0	− 10
□	□	□	□

10	11	11	9
− 9	− 11	− 6	− 1
□	□	□	□

10	10	11
− 10	− 4	− 3
□	□	□

9	11	10
− 8	− 2	− 1
□	□	□

9	9	11	8	8	9	8
− 4	− 0	− 8	− 8	− 5	− 6	− 1
□	□	□	□	□	□	□

10	8	8	12	11	12
− 5	− 7	− 3	− 7	− 5	− 5
□	□	□	□	□	□

10

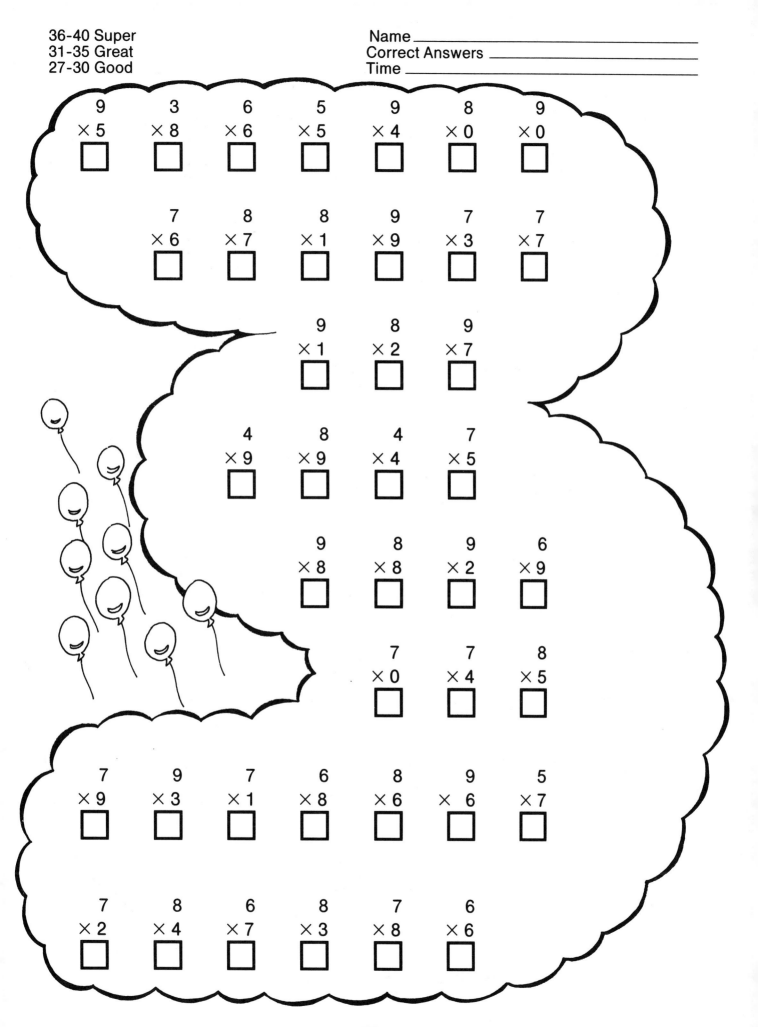

36-40 Super
31-35 Great
27-30 Good

Name _____
Correct Answers _____
Time _____

$\begin{array}{r}9\\ \times 5\\ \hline \square\end{array}$ $\begin{array}{r}3\\ \times 8\\ \hline \square\end{array}$ $\begin{array}{r}6\\ \times 6\\ \hline \square\end{array}$ $\begin{array}{r}5\\ \times 5\\ \hline \square\end{array}$ $\begin{array}{r}9\\ \times 4\\ \hline \square\end{array}$ $\begin{array}{r}8\\ \times 0\\ \hline \square\end{array}$ $\begin{array}{r}9\\ \times 0\\ \hline \square\end{array}$

$\begin{array}{r}7\\ \times 6\\ \hline \square\end{array}$ $\begin{array}{r}8\\ \times 7\\ \hline \square\end{array}$ $\begin{array}{r}8\\ \times 1\\ \hline \square\end{array}$ $\begin{array}{r}9\\ \times 9\\ \hline \square\end{array}$ $\begin{array}{r}7\\ \times 3\\ \hline \square\end{array}$ $\begin{array}{r}7\\ \times 7\\ \hline \square\end{array}$

$\begin{array}{r}9\\ \times 1\\ \hline \square\end{array}$ $\begin{array}{r}8\\ \times 2\\ \hline \square\end{array}$ $\begin{array}{r}9\\ \times 7\\ \hline \square\end{array}$

$\begin{array}{r}4\\ \times 9\\ \hline \square\end{array}$ $\begin{array}{r}8\\ \times 9\\ \hline \square\end{array}$ $\begin{array}{r}4\\ \times 4\\ \hline \square\end{array}$ $\begin{array}{r}7\\ \times 5\\ \hline \square\end{array}$

$\begin{array}{r}9\\ \times 8\\ \hline \square\end{array}$ $\begin{array}{r}8\\ \times 8\\ \hline \square\end{array}$ $\begin{array}{r}9\\ \times 2\\ \hline \square\end{array}$ $\begin{array}{r}6\\ \times 9\\ \hline \square\end{array}$

$\begin{array}{r}7\\ \times 0\\ \hline \square\end{array}$ $\begin{array}{r}7\\ \times 4\\ \hline \square\end{array}$ $\begin{array}{r}8\\ \times 5\\ \hline \square\end{array}$

$\begin{array}{r}7\\ \times 9\\ \hline \square\end{array}$ $\begin{array}{r}9\\ \times 3\\ \hline \square\end{array}$ $\begin{array}{r}7\\ \times 1\\ \hline \square\end{array}$ $\begin{array}{r}6\\ \times 8\\ \hline \square\end{array}$ $\begin{array}{r}8\\ \times 6\\ \hline \square\end{array}$ $\begin{array}{r}9\\ \times 6\\ \hline \square\end{array}$ $\begin{array}{r}5\\ \times 7\\ \hline \square\end{array}$

$\begin{array}{r}7\\ \times 2\\ \hline \square\end{array}$ $\begin{array}{r}8\\ \times 4\\ \hline \square\end{array}$ $\begin{array}{r}6\\ \times 7\\ \hline \square\end{array}$ $\begin{array}{r}8\\ \times 3\\ \hline \square\end{array}$ $\begin{array}{r}7\\ \times 8\\ \hline \square\end{array}$ $\begin{array}{r}6\\ \times 6\\ \hline \square\end{array}$

11

GA1392

Name _____
Correct Answers _____
Time _____

☐ $3\overline{)12}$ ☐ $5\overline{)15}$ ☐ $4\overline{)12}$ ☐ $2\overline{)6}$ ☐ $10\overline{)20}$ ☐ $3\overline{)15}$ ☐ $6\overline{)18}$

☐ $2\overline{)14}$ ☐ $5\overline{)10}$ ☐ $2\overline{)16}$ ☐ $3\overline{)18}$ ☐ $1\overline{)10}$ ☐ $7\overline{)14}$

☐ $9\overline{)18}$ ☐ $3\overline{)21}$ ☐ $2\overline{)30}$

☐ $2\overline{)4}$ ☐ $2\overline{)18}$ ☐ $1\overline{)11}$ ☐ $6\overline{)30}$

☐ $9\overline{)27}$ ☐ $5\overline{)20}$ ☐ $6\overline{)24}$

☐ $7\overline{)21}$ ☐ $3\overline{)9}$ ☐ $2\overline{)10}$

☐ $10\overline{)30}$ ☐ $1\overline{)3}$ ☐ $3\overline{)30}$ ☐ $2\overline{)20}$ ☐ $1\overline{)15}$ ☐ $3\overline{)27}$ ☐ $5\overline{)30}$

☐ $7\overline{)28}$ ☐ $4\overline{)20}$ ☐ $8\overline{)24}$ ☐ $4\overline{)28}$ ☐ $5\overline{)25}$ ☐ $2\overline{)24}$ ☐ $1\overline{)5}$

12

GA1392

36-40 Super
31-35 Great
27-30 Good

Name _____
Correct Answers _____
Time _____

13

GA1392

36-40 Super
31-35 Great
27-30 Good

Name _____
Correct Answers _____
Time _____

This page is for you to create your own daily drill.

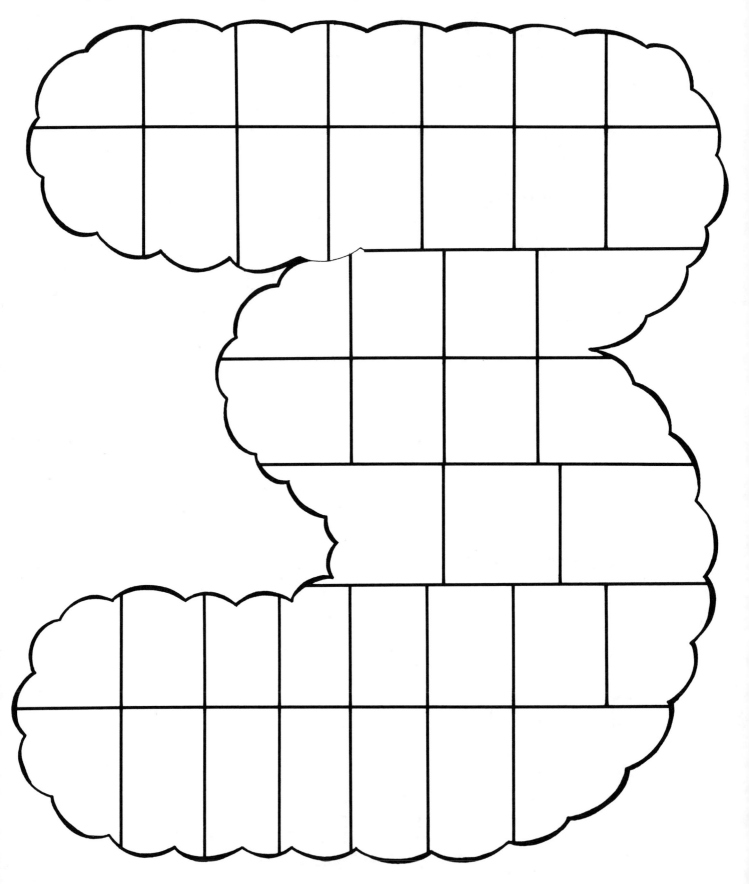

GA1392

36-40 Super
31-35 Great
27-30 Good

Name _____
Correct Answers _____
Time _____

Row 1:
7	9	3	9	6	5
+ 5	+ 1	+ 8	+ 9	+ 7	+ 9
□	□	□	□	□	□

Row 2:
4	11	10	7	8	4
+ 8	+ 9	+ 2	+ 7	+ 5	+ 10
□	□	□	□	□	□

Row 3:
7	6	5
+ 8	+ 12	+ 8
□	□	□

Row 4:
5	6	6
+ 10	+ 6	+ 5
□	□	□

Row 5:
10	9	5
+ 0	+ 10	+ 0
□	□	□

Row 6:
4	4	7	3
+ 9	+ 7	+ 10	+ 9
□	□	□	□

Row 7:
8	3	8	6	8	8	8
+ 9	+ 10	+ 8	+ 10	+ 0	+ 10	+ 2
□	□	□	□	□	□	□

Row 8:
7	1	7	6	4	10	11	12
+ 9	+ 10	+ 4	+ 3	+ 4	+ 10	+ 0	+ 2
□	□	□	□	□	□	□	□

15

Name _____
Correct Answers _____
Time _____

12	10	11	9	12	10
− 4	− 2	− 9	− 9	− 11	− 6
□	□	□	□	□	□

11	10	12	12	9	11	9
− 0	− 7	− 10	− 9	− 7	− 1	− 2
□	□	□	□	□	□	□

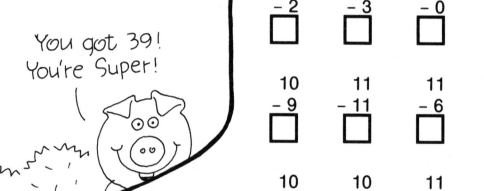

You got 39!
You're Super!

12	10	12	11
− 2	− 3	− 0	− 10
□	□	□	□

10	11	11	9
− 9	− 11	− 6	− 1
□	□	□	□

10	10	11	9
− 10	− 4	− 3	− 8
□	□	□	□

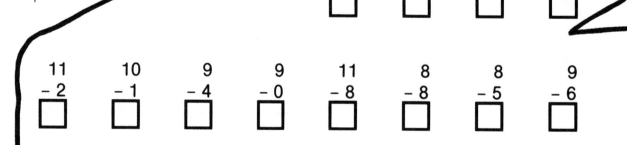

11	10	9	9	11	8	8	9
− 2	− 1	− 4	− 0	− 8	− 8	− 5	− 6
□	□	□	□	□	□	□	□

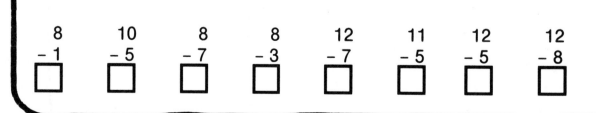

8	10	8	8	12	11	12	12
− 1	− 5	− 7	− 3	− 7	− 5	− 5	− 8
□	□	□	□	□	□	□	□

16

Name _____
Correct Answers _____
Time _____

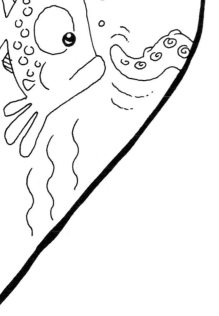

```
   9       3       6       5       9       8       9
 × 5     × 8     × 6     × 5     × 4     × 0     × 0
 [  ]    [  ]    [  ]    [  ]    [  ]    [  ]    [  ]
```

```
           7       8       8       9       7       7
         × 6     × 7     × 1     × 9     × 3     × 7
         [  ]    [  ]    [  ]    [  ]    [  ]    [  ]
```

```
                           9       8       9
                         × 1     × 2     × 7
                         [  ]    [  ]    [  ]
```

```
                   4       8       4       7
                 × 9     × 9     × 4     × 5
                 [  ]    [  ]    [  ]    [  ]
```

```
                           9       8       9
                         × 8     × 8     × 2
                         [  ]    [  ]    [  ]
```

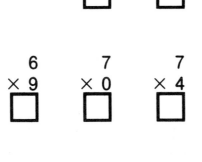

```
                   6       7       7
                 × 9     × 0     × 4
                 [  ]    [  ]    [  ]
```

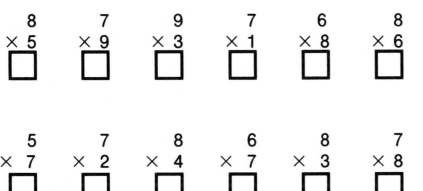

```
   8       7       9       7       6       8       9
 × 5     × 9     × 3     × 1     × 8     × 6     × 6
 [  ]    [  ]    [  ]    [  ]    [  ]    [  ]    [  ]
```

```
   5       7       8       6       8       7       6
 × 7     × 2     × 4     × 7     × 3     × 8     × 6
 [  ]    [  ]    [  ]    [  ]    [  ]    [  ]    [  ]
```

17

Name _____
Correct Answers _____
Time _____

□ 3⟌12   □ 5⟌15   □ 4⟌12   □ 2⟌6   □ 10⟌20   □ 3⟌15   □ 6⟌18

□ 2⟌14   □ 5⟌10   □ 2⟌16   □ 3⟌18   □ 1⟌10   □ 7⟌14

□ 9⟌18   □ 3⟌21   □ 2⟌30

□ 2⟌4   □ 2⟌18   □ 1⟌11   □ 6⟌30

□ 9⟌27   □ 5⟌20   □ 6⟌24

□ 10⟌30   □ 1⟌13   □ 3⟌30

□ 2⟌20   □ 1⟌15   □ 3⟌27   □ 5⟌30   □ 7⟌28   □ 4⟌20   □ 8⟌24

□ 4⟌28   □ 5⟌25   □ 2⟌24   □ 1⟌5   □ 7⟌21   □ 3⟌9   □ 2⟌10

18

36-40 Super
31-35 Great
27-30 Good

Name _____
Correct Answers _____
Time _____

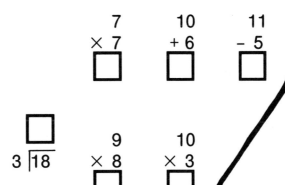

7⟌21 □

9
× 4
□

8
+ 9
□

12
+ 0
□

12
− 8
□

10
− 5
□

6
× 6
□

□
3⟌15

7
× 9
□

□
2⟌14

□
4⟌36

12
− 0
□

9
× 9
□

□
5⟌15

12
− 4
□

7⟌49 □

3
+ 12
□

□
8⟌16

8
× 10
□

4⟌4 □

7
× 7
□

10
+ 6
□

11
− 5
□

□
3⟌18

9
× 8
□

10
× 3
□

2
× 10
□

11
− 11
□

12
+ 0
□

6
× 8
□

4
+ 9
□

9⟌18 □

9
× 7
□

□
9⟌63

11
− 1
□

□
24⟌24

3
× 8
□

10
× 10
□

4
× 6
□

7⟌28 □

Copyright © 1992, Good Apple

19

GA1392

36-40 Super
31-35 Great
27-30 Good

Name _____

Correct Answers _____

Time _____

This page is for you to create your own daily drill.

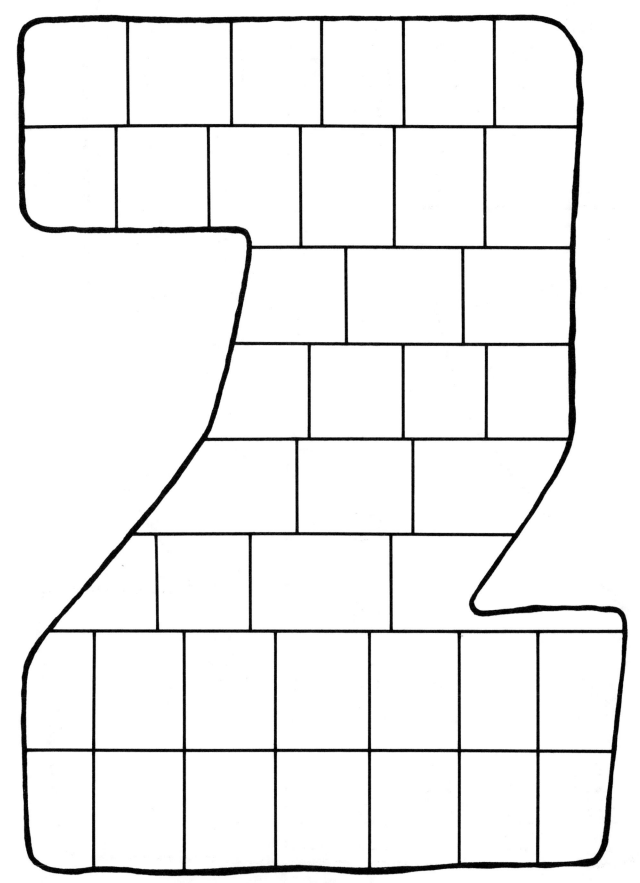

GA1392

Name _____
Correct Answers _____
Time _____

7 + 5 ☐	9 + 1 ☐	3 + 8 ☐	9 + 9 ☐	6 + 7 ☐
5 + 9 ☐	4 + 8 ☐	11 + 9 ☐	10 + 2 ☐	7 + 7 ☐
8 + 5 ☐	4 + 10 ☐	7 + 8 ☐	6 + 12 ☐	5 + 8 ☐
5 + 10 ☐	6 + 6 ☐	6 + 5 ☐	10 + 0 ☐	9 + 10 ☐
5 + 0 ☐	4 + 9 ☐	4 + 7 ☐	7 + 10 ☐	3 + 9 ☐
8 + 9 ☐	3 + 10 ☐	8 + 8 ☐	6 + 10 ☐	8 + 0 ☐
8 + 10 ☐	8 + 2 ☐	7 + 9 ☐	1 + 10 ☐	7 + 4 ☐
6 + 3 ☐	4 + 4 ☐	10 + 10 ☐	11 + 0 ☐	12 + 2 ☐

c'mon, give it a try! it's easy!

21

GA1392

Name _____
Correct Answers _____
Time _____

12 − 4 □	10 − 2 □	11 − 9 □	9 − 9 □	12 − 11 □
10 − 6 □	11 − 0 □	10 − 7 □	12 − 10 □	12 − 9 □
9 − 7 □	11 − 1 □	9 − 2 □	12 − 2 □	10 − 3 □
12 − 0 □	11 − 10 □	10 − 9 □	11 − 11 □	11 − 6 □
9 − 1 □	10 − 10 □	10 − 4 □	11 − 3 □	9 − 8 □
11 − 2 □	10 − 1 □	9 − 4 □	9 − 0 □	11 − 8 □
8 − 8 □	8 − 5 □	9 − 6 □	8 − 1 □	10 − 5 □
8 − 7 □	8 − 3 □	12 − 7 □	11 − 5 □	12 − 5 □

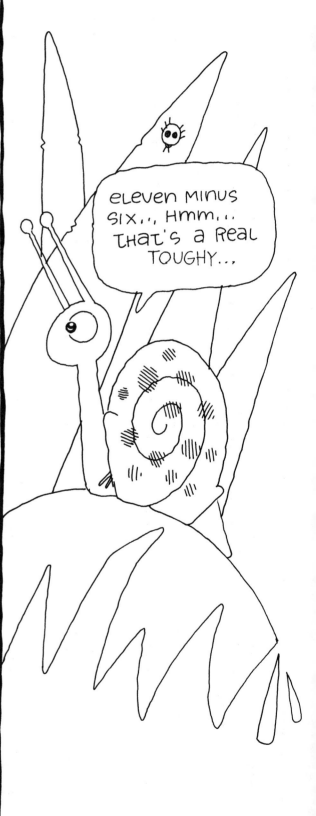

eLeven minus six... Hmm... THaT's a ReaL TOUGHy...

22

Name _____
Correct Answers _____
Time _____

I'm READY!

9 × 5 ☐	3 × 8 ☐	6 × 6 ☐	5 × 5 ☐	9 × 4 ☐
8 × 0 ☐	9 × 0 ☐	7 × 6 ☐	8 × 7 ☐	8 × 1 ☐
9 × 9 ☐	7 × 3 ☐	7 × 7 ☐	9 × 1 ☐	8 × 2 ☐
9 × 7 ☐	4 × 9 ☐	8 × 9 ☐	4 × 4 ☐	7 × 5 ☐
9 × 8 ☐	8 × 8 ☐	9 × 2 ☐	6 × 9 ☐	7 × 0 ☐
7 × 4 ☐	8 × 5 ☐	7 × 9 ☐	9 × 3 ☐	7 × 1 ☐
6 × 8 ☐	8 × 6 ☐	9 × 6 ☐	5 × 7 ☐	7 × 2 ☐
8 × 4 ☐	6 × 7 ☐	8 × 3 ☐	7 × 8 ☐	6 × 6 ☐

GA1392

Name _____
Correct Answers _____
Time _____

□
3√12   5√15   4√12   2√6   10√20

□
3√15   6√18   2√14   5√10   2√16

□
3√18   1√10   7√14   9√18   3√21

□
2√30   2√4   2√18   1√11   6√30

□
9√27   5√20   6√24   10√30   1√13

□
3√30   2√20   1√15   3√27   5√30

□
7√28   4√20   8√24   4√28   5√25

□
2√24   1√5   7√21   3√9   2√10

TRUST Me...
THIS IS Gonna
Be a Piece OF
cake!

24

Name _____
Correct Answers _____
Time _____

$7\overline{)21}$ □

9×4 □

$8 + 9$ □

$12 + 0$ □

$12 - 8$ □

$10 - 5$ □

6×6 □

□ $3\overline{)15}$

7×9 □

□ $2\overline{)14}$

□ $4\overline{)36}$

$12 - 0$ □

9×9 □

$5\overline{)15}$ □

$12 - 4$ □

□ $7\overline{)49}$

$3 + 12$ □

$8\overline{)16}$ □

$8 + 10$ □

$4\overline{)4}$ □

7×7 □

$10 + 6$ □

$11 - 5$ □

□ $3\overline{)18}$

9×8 □

10×3 □

2×10 □

$11 - 11$ □

$12 - 0$ □

6×8 □

$4 + 9$ □

□ $9\overline{)18}$

9×7 □

□ $9\overline{)63}$

$11 - 1$ □

□ $24\overline{)24}$

3×8 □

10×10 □

4×6 □

$7\overline{)28}$ □

IF I can
DO IT,
anyone
can!

25

36-40 Super
31-35 Great
27-30 Good

Name _____

Correct Answers _____

Time _____

This page is for you to create your own daily drill.

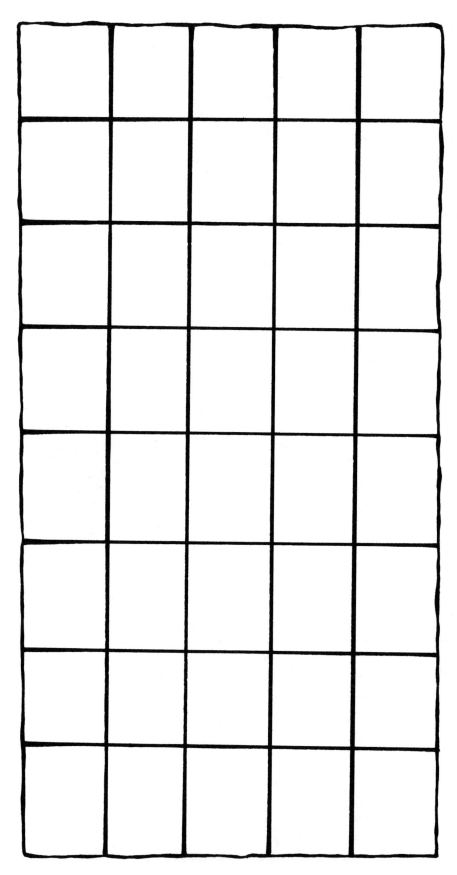

GA1392

Let's Talk Math

Let's Talk Math is a math facts drill that means exactly what it says. It means to talk the language of math. It means to think the language of math. It means to break down barriers that interfere with the understanding and enjoyment of math.

Rather than just the acknowledgement of math symbols that probably do not have much meaning to many students, Let's Talk Math tries to place meaning and encourage understanding when students work with simple or even difficult mathematical sentences.

This drill attempts to have students think about the addition symbol (+) in the same light as its counterpart, the word *plus*. Consider the symbol for subtraction (–) as the word *minus*, 9 as *nine*, 14 as *fourteen* and the symbol = as the word *equals*. You will see a tie-in with math symbols and math language. For example, a student might see the equation

$$15 \div 5 = \square$$

In Let's Talk Math this basic fact might be written as

$$15 \text{ divided by } 5 = \square$$

or

$$\text{fifteen} \div \text{five equals } \square$$

It is hoped that a better student understanding of basic math facts will alleviate math phobia.

GA1392

Let's Talk Math 1    Name_____

Addition to 12

five + 4 = ☐	three + 9 = ☐
3 plus 6 = ☐	4 + 4 equals ☐
7 + one = ☐	eight plus 2 = ☐
1 + nine = ☐	two + five = ☐
4 + 6 equals ☐	nine + one = ☐
eight + 2 = ☐	seven plus 3 = ☐
five + six = ☐	one + 7 = ☐
8 plus 0 = ☐	10 plus 2 = ☐
four + eight = ☐	six + 6 equals ☐
5 plus five = ☐	6 plus 4 = ☐

Let's Talk Math 2    Name _____

Addition to 18

7 + five = ☐	12 plus 4 = ☐
three + 9 = ☐	11 plus seven = ☐
five plus 8 = ☐	five + 9 = ☐
7 plus seven = ☐	16 plus two = ☐
8 + four = ☐	7 plus 8 equals ☐
nine + 7 = ☐	ten + four = ☐
eight plus 8 = ☐	3 plus 11 = ☐
9 + nine = ☐	5 + eight = ☐
ten + 8 = ☐	ten + one = ☐
four plus 4 = ☐	2 + 12 equals ☐

                                        GA1392

Name_____

Addition to 30

twenty plus 4 = ☐

14 plus 2 = ☐

fifteen + 15 = ☐

16 plus 4 = ☐

4 plus twenty-two = ☐

eighteen plus 10 = ☐

eleven plus 10 = ☐

21 + nine = ☐

seventeen + 3 = ☐

sixteen plus 10 = ☐

7 plus 23 = ☐

20 + eight = ☐

5 plus 25 = ☐

fifteen + 5 = ☐

20 plus ten = ☐

13 + thirteen = ☐

18 plus 10 = ☐

17 plus ten = ☐

12 + twelve = ☐

eight + twelve = ☐

GA1392

Let's Talk Math 4

Name_____

Subtraction from 12

11 – five = ☐	12 minus 12 = ☐
10 – six = ☐	ten – 9 = ☐
12 – 0 equals ☐	6 – six = ☐
8 – five = ☐	12 – ten = ☐
seven – four = ☐	6 – five ☐
ten – zero = ☐	11 – four = ☐
9 – 8 equals ☐	12 minus five = ☐
7 – six = ☐	10 – four = ☐
10 – three = ☐	4 – four equals ☐
9 – seven = ☐	8 – six = ☐

GA1392

Name_____

Subtraction from 18

sixteen – 10 = ☐	17 – three = ☐
18 – nine = ☐	15 – 14 equals ☐
fifteen – 5 = ☐	18 – one = ☐
18 – eight = ☐	16 minus 3 = ☐
sixteen – four = ☐	13 – eight = ☐
15 – ten = ☐	17 – five = ☐
17 – seven = ☐	16 minus five = ☐
16 – one = ☐	18 minus 2 = ☐
14 minus 2 equals ☐	14 minus 4 = ☐
13 – ten = ☐	12 – nine = ☐

GA1392

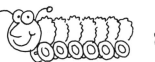
Subtraction from 24—no regrouping

twenty-three – 3 = ☐	eighteen minus 8 = ☐
17 minus five = ☐	sixteen – zero = ☐
fifteen – three = ☐	19 – five = ☐
nineteen minus 2 = ☐	15 minus 5 = ☐
16 – four = ☐	24 minus 2 = ☐
23 – zero = ☐	sixteen – 6 = ☐
19 – ten = ☐	twenty-four – 4 = ☐
seventeen – 4 = ☐	18 – ten = ☐
24 minus 3 equals ☐	24 minus 0 = ☐
15 minus four = ☐	17 – six = ☐

GA1392

Name _____

Multiplication to 20

3 times five = ☐	3 times 3 = ☐
four × two = ☐	2 × ten = ☐
1 × eight equals ☐	8 times 0 = ☐
five times 4 = ☐	three × 6 = ☐
four × 4 = ☐	10 × 1 equals ☐
9 × two = ☐	5 × three = ☐
two × eight = ☐	4 × 5 equals ☐
six times 3 = ☐	2 times 4 = ☐
2 times 2 = ☐	3 times four = ☐
10 × two = ☐	2 times 9 = ☐

GA1392

Name_____

Multiplication to 30

six times 4 = ☐	4 × five = ☐
8 times 3 = ☐	3 × three = ☐
3 × nine = ☐	1 × 8 equals ☐
10 × two = ☐	3 × eight = ☐
2 times 9 = ☐	10 times 3 = ☐
five × 5 = ☐	1 × 1 equals ☐
7 × three = ☐	9 × three = ☐
2 times 2 = ☐	4 × 4 equals ☐
8 × two = ☐	5 × 3 equals ☐
3 × ten = ☐	9 times two = ☐

GA1392

Name_____

Multiplication to 50

three × six = ☐

10 × two = ☐

7 × seven = ☐

6 times 3 = ☐

4 × nine = ☐

ten × five = ☐

five times 8 = ☐

4 times 10 = ☐

6 × seven = ☐

8 × six = ☐

8 × five = ☐

4 × three = ☐

6 × eight = ☐

nine × four = ☐

2 × 10 equals ☐

seven × six = ☐

5 × nine = ☐

3 × nine = ☐

9 × five = ☐

3 × 4 equals ☐

36

GA1392

Name_____

Division to 25

18 ÷ three = ☐	ten ÷ 5 = ☐
12 divided by 6 = ☐	14 ÷ two = ☐
fifteen ÷ 5 = ☐	24 divided by 6 = ☐
sixteen ÷ two = ☐	15 divided by 3 = ☐
12 ÷ four = ☐	12 ÷ three = ☐
10 ÷ two = ☐	18 ÷ six = ☐
24 ÷ eight = ☐	12 ÷ two = ☐
ten ÷ ten = ☐	24 ÷ three = ☐
14 ÷ seven = ☐	21 ÷ seven = ☐
18 ÷ nine = ☐	25 ÷ five = ☐

GA1392

Let's Talk Math 11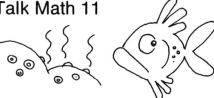

Name_____

Division to 40

30 ÷ ten = ☐	35 ÷ seven = ☐
25 divided by 5 = ☐	28 ÷ four = ☐
forty ÷ 5 = ☐	36 ÷ four = ☐
30 divided by 3 = ☐	24 ÷ two = ☐
36 ÷ nine = ☐	30 divided by 5 = ☐
30 ÷ six = ☐	24 ÷ 1 equals ☐
24 ÷ 8 equals ☐	40 ÷ 8 equals ☐
36 ÷ 6 equals ☐	30 ÷ two = ☐
24 ÷ three = ☐	28 ÷ seven = ☐
35 ÷ 5 equals ☐	20 ÷ ten = ☐

Copyright © 1992, Good Apple

GA1392

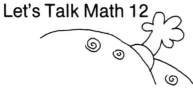
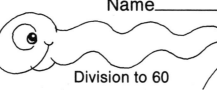
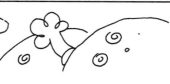

Name_____

Division to 60

54 divided by 9 = ☐

forty ÷ 5 = ☐

48 ÷ eight = ☐

36 ÷ nine = ☐

sixty ÷ ten = ☐

forty ÷ ten = ☐

fifty ÷ 10 = ☐

forty-two ÷ 7 = ☐

48 ÷ 6 = ☐

36 ÷ six = ☐

49 ÷ 7 equals ☐

42 ÷ six = ☐

forty-five ÷ 5 = ☐

54 ÷ six = ☐

fifty ÷ 5 = ☐

36 ÷ four = ☐

45 ÷ 9 equals ☐

44 ÷ two = ☐

40 ÷ two = ☐

forty ÷ 8 = ☐

Name_____

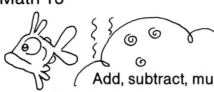

Add, subtract, multiply or divide facts to 20

ten × 2 = ☐

12 × zero = ☐

twenty minus 10 = ☐

8 plus eight = ☐

15 minus 4 = ☐

twenty ÷ four = ☐

8 minus eight = ☐

14 + four = ☐

two times 6 = ☐

11 plus 8 = ☐

12 divided by 3 = ☐

3 × one = ☐

fourteen − 3 = ☐

7 times 2 = ☐

twenty times 0 = ☐

12 plus 6 = ☐

20 ÷ two = ☐

16 minus 5 = ☐

four + 5 = ☐

6 + thirteen = ☐

GA1392

Name _____

Add, subtract, multiply or divide facts to 30

10 + ten = ☐	18 − six = ☐
twenty ÷ two = ☐	3 × ten = ☐
nine plus 8 = ☐	10 ÷ ten = ☐
six × 5 = ☐	14 − three = ☐
ten ÷ five = ☐	8 plus ten = ☐
14 − nine = ☐	twenty ÷ 5 = ☐
five × 4 = ☐	3 times nine = ☐
16 ÷ four = ☐	4 × 5 equals ☐
15 minus 6 = ☐	16 + four = ☐
12 × 2 equals ☐	18 plus 6 = ☐

41   GA1392

Let's Talk Math 15  Name_____

Add, subtract, multiply or divide facts to 40

8 times five = ☐	7 times five = ☐
18 minus 4 = ☐	21 minus 3 = ☐
forty ÷ 8 = ☐	36 ÷ 3 equals ☐
20 plus 20 = ☐	12 × two = ☐
16 – ten = ☐	40 ÷ ten = ☐
36 ÷ 6 equals ☐	30 plus 8 = ☐
twenty-four – 4 = ☐	20 – five = ☐
10 × four = ☐	28 ÷ two = ☐
15 + ten = ☐	22 plus 8 = ☐
25 + four = ☐	15 × two = ☐

GA1392

Name_____

Add, subtract, multiply or divide facts to 50

12 times 2 = ☐	32 + seven = ☐
20 times 2 = ☐	5 × ten = ☐
eight × six = ☐	twenty-five × 2 = ☐
35 divided by 7 = ☐	38 minus 5 = ☐
42 plus eight = ☐	49 ÷ 7 equals ☐
ten × five = ☐	24 ÷ three = ☐
7 times seven = ☐	35 minus 5 = ☐
15 × two = ☐	31 minus 0 = ☐
50 divided by 5 = ☐	six times 5 = ☐
42 plus 7 = ☐	37 plus 10 = ☐

GA1392

Name_____

Add, subtract, multiply or divide facts to 60

sixteen times 3 = ☐	nine times 5 = ☐
seventeen × zero = ☐	twelve × three = ☐
51 plus eight = ☐	57 minus fourteen = ☐
zero × forty-one = ☐	twenty plus 29 = ☐
42 divided by 3 = ☐	42 minus thirteen = ☐
eight times 7 = ☐	fifteen + twenty-five = ☐
59 – fifty-one = ☐	17 plus twenty-one = ☐
fifteen × four = ☐	thirteen plus 29 = ☐
3 times eighteen = ☐	60 minus eighteen = ☐
60 divided by 10 = ☐	49 ÷ seven equals ☐

Your Guess Is as Good as Mine

This drill requires students to immediately consider the three different possible answers to each problem (four answers when using addition, subtraction, multiplication or division). This drill is not looking for one response from students, such as

$$
\begin{array}{r}
8 \\
\times\ 2 \\
\hline
16
\end{array}
$$

The drill encourages more thinking on the part of each student. For example, 8: If there's an 8 with a 2, the student now wonders will the

$$\underline{2}$$

teacher add $8 + 2 = 10$, subtract $8 - 2 = 6$, multiply $8 \times 2 = 16$ or divide $8 \div 2 = 4$.

Again class points are scored when there is a match in answers between the student who is called upon to respond and the teacher.

There are twenty-five problems on every page of this drill. There is a separate answer sheet that will correspond with each drill page. It is up to the individual student or team of students to select an operation sign from addition, subtraction or multiplication. Once a sign has been selected, apply the particular sign to the problem and solve it. The same operational sign will not be used for all of the problems.

For example: Suppose this were part of the drill sheet.

a. 9	b. 7	c. 3	d. 9	e. 4
5	6	10	9	7

From the operations of addition, subtraction and multiplication, suppose the student chose the following. His/Her drill sheet might look something like this:

a. 9	b. 7	c. 3	d. 9	e. 4
+ 5	× 6	× 10	– 9	+ 7
14	42	30	0	11

Upon examination of the answer sheet, the student might find this:

a. 9	b. 7	c. 3	d. 9	e. 4
– 5	+ 6	× 10	× 9	+ 7
4	13	30	81	11

In this sample the student would score 2 points.

1 point for matching problem c. 3 and 1 point for matching problem
e. 4   × 10
   + 7   ———
   ——   30
    11

Participants should follow this procedure for scoring:

1. Solve the problems on the drill sheet by selecting one operation sign from addition, subtraction or multiplication.

2. After completing the twenty problems, check the corresponding answer sheet.

3. One point is given to each problem in which the operation sign and answer are the same.

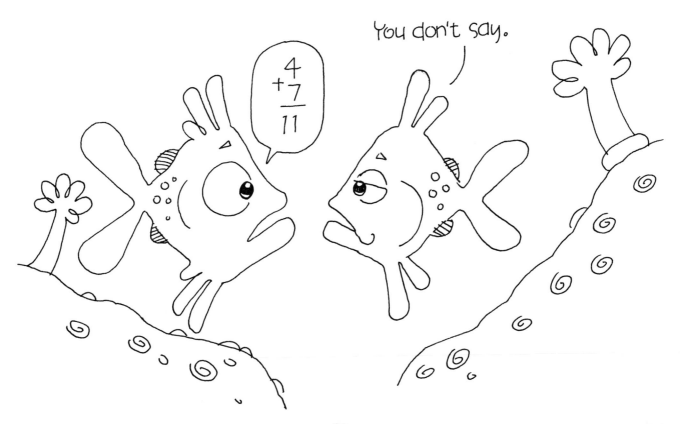

GA1392

Drill Sheet 1

a. 5 3	b. 6 5	c. 9 7	d. 2 4	e. 8 8
f. 7 3	g. 10 2	h. 4 9	i. 6 7	j. 3 6
k. 5 5	l. 9 0	m. 4 4	n. 0 8	o. 9 2
p. 10 3	q. 4 5	r. 7 6	s. 3 7	t. 8 4
u. 6 6	v. 8 3	w. 9 3	x. 10 10	y. 7 7

Name _____

47

GA1392

Answer Sheet 1

a. $\begin{array}{r} 5 \\ -\ 3 \\ \hline 2 \end{array}$	**b.** $\begin{array}{r} 6 \\ \times\ 5 \\ \hline 30 \end{array}$	**c.** $\begin{array}{r} 9 \\ -\ 7 \\ \hline 2 \end{array}$	**d.** $\begin{array}{r} 2 \\ +\ 4 \\ \hline 6 \end{array}$	**e.** $\begin{array}{r} 8 \\ -\ 8 \\ \hline 0 \end{array}$
f. $\begin{array}{r} 7 \\ \times\ 3 \\ \hline 21 \end{array}$	**g.** $\begin{array}{r} 10 \\ +\ 2 \\ \hline 12 \end{array}$	**h.** $\begin{array}{r} 4 \\ +\ 9 \\ \hline 13 \end{array}$	**i.** $\begin{array}{r} 6 \\ \times\ 7 \\ \hline 42 \end{array}$	**j.** $\begin{array}{r} 3 \\ \times\ 6 \\ \hline 18 \end{array}$
k. $\begin{array}{r} 5 \\ +\ 5 \\ \hline 10 \end{array}$	**l.** $\begin{array}{r} 9 \\ \times\ 0 \\ \hline 0 \end{array}$	**m.** $\begin{array}{r} 4 \\ \times\ 4 \\ \hline 16 \end{array}$	**n.** $\begin{array}{r} 0 \\ +\ 8 \\ \hline 8 \end{array}$	**o.** $\begin{array}{r} 9 \\ +\ 2 \\ \hline 11 \end{array}$
p. $\begin{array}{r} 10 \\ \times\ 3 \\ \hline 30 \end{array}$	**q.** $\begin{array}{r} 4 \\ +\ 5 \\ \hline 9 \end{array}$	**r.** $\begin{array}{r} 7 \\ +\ 6 \\ \hline 13 \end{array}$	**s.** $\begin{array}{r} 3 \\ \times\ 7 \\ \hline 21 \end{array}$	**t.** $\begin{array}{r} 8 \\ +\ 4 \\ \hline 12 \end{array}$
u. $\begin{array}{r} 6 \\ \times\ 6 \\ \hline 36 \end{array}$	**v.** $\begin{array}{r} 8 \\ -\ 3 \\ \hline 5 \end{array}$	**w.** $\begin{array}{r} 9 \\ -\ 3 \\ \hline 6 \end{array}$	**x.** $\begin{array}{r} 10 \\ \times\ 10 \\ \hline 100 \end{array}$	**y.** $\begin{array}{r} 7 \\ -\ 7 \\ \hline 0 \end{array}$

LET'S TALK MATH!

Name _____

48

GA1392

Drill Sheet 2

a. 9 3	b. 5 6	c. 8 5	d. 1 1	e. 7 2
f. 6 6	g. 10 9	h. 0 10	i. 3 5	j. 9 9
k. 10 10	l. 5 5	m. 2 7	n. 10 1	o. 6 4
p. 9 2	q. 3 8	r. 8 10	s. 3 3	t. 0 5
u. 5 9	v. 10 7	w. 8 8	x. 5 4	y. 7 6

Name _____

GA1392

Answer Sheet 2

a.	b.	c.	d.	e.
$$\begin{array}{r} 9 \\ +\ 3 \\ \hline 12 \end{array}$$	$$\begin{array}{r} 5 \\ +\ 6 \\ \hline 11 \end{array}$$	$$\begin{array}{r} 8 \\ \times\ 5 \\ \hline 40 \end{array}$$	$$\begin{array}{r} 1 \\ \times\ 1 \\ \hline 1 \end{array}$$	$$\begin{array}{r} 7 \\ \times\ 2 \\ \hline 14 \end{array}$$
f.	g.	h.	i.	j.
$$\begin{array}{r} 6 \\ -\ 6 \\ \hline 0 \end{array}$$	$$\begin{array}{r} 10 \\ \times\ 9 \\ \hline 90 \end{array}$$	$$\begin{array}{r} 0 \\ \times\ 10 \\ \hline 0 \end{array}$$	$$\begin{array}{r} 3 \\ \times\ 5 \\ \hline 15 \end{array}$$	$$\begin{array}{r} 9 \\ \times\ 9 \\ \hline 81 \end{array}$$
k.	l.	m.	n.	o.
$$\begin{array}{r} 10 \\ +\ 10 \\ \hline 20 \end{array}$$	$$\begin{array}{r} 5 \\ \times\ 5 \\ \hline 25 \end{array}$$	$$\begin{array}{r} 2 \\ +\ 7 \\ \hline 9 \end{array}$$	$$\begin{array}{r} 10 \\ \times\ 1 \\ \hline 10 \end{array}$$	$$\begin{array}{r} 6 \\ -\ 4 \\ \hline 2 \end{array}$$
p.	q.	r.	s.	t.
$$\begin{array}{r} 9 \\ +\ 2 \\ \hline 11 \end{array}$$	$$\begin{array}{r} 3 \\ \times\ 8 \\ \hline 24 \end{array}$$	$$\begin{array}{r} 8 \\ \times\ 10 \\ \hline 80 \end{array}$$	$$\begin{array}{r} 3 \\ -\ 3 \\ \hline 0 \end{array}$$	$$\begin{array}{r} 0 \\ \times\ 5 \\ \hline 0 \end{array}$$
u.	v.	w.	x.	y.
$$\begin{array}{r} 5 \\ \times\ 9 \\ \hline 45 \end{array}$$	$$\begin{array}{r} 10 \\ \times\ 7 \\ \hline 70 \end{array}$$	$$\begin{array}{r} 8 \\ +\ 8 \\ \hline 16 \end{array}$$	$$\begin{array}{r} 5 \\ -\ 4 \\ \hline 1 \end{array}$$	$$\begin{array}{r} 7 \\ +\ 6 \\ \hline 13 \end{array}$$

Name _____

50

GA1392

Drill Sheet 3

a. $\begin{array}{r} 5 \\ 8 \\ \hline \end{array}$	b. $\begin{array}{r} 7 \\ 7 \\ \hline \end{array}$	c. $\begin{array}{r} 6 \\ 3 \\ \hline \end{array}$	d. $\begin{array}{r} 9 \\ 1 \\ \hline \end{array}$	e. $\begin{array}{r} 4 \\ 10 \\ \hline \end{array}$
f. $\begin{array}{r} 3 \\ 2 \\ \hline \end{array}$	g. $\begin{array}{r} 10 \\ 7 \\ \hline \end{array}$	h. $\begin{array}{r} 8 \\ 8 \\ \hline \end{array}$	i. $\begin{array}{r} 6 \\ 7 \\ \hline \end{array}$	j. $\begin{array}{r} 4 \\ 9 \\ \hline \end{array}$
k. $\begin{array}{r} 8 \\ 10 \\ \hline \end{array}$	l. $\begin{array}{r} 1 \\ 0 \\ \hline \end{array}$	m. $\begin{array}{r} 10 \\ 10 \\ \hline \end{array}$	n. $\begin{array}{r} 8 \\ 9 \\ \hline \end{array}$	o. $\begin{array}{r} 7 \\ 1 \\ \hline \end{array}$
p. $\begin{array}{r} 3 \\ 5 \\ \hline \end{array}$	q. $\begin{array}{r} 10 \\ 3 \\ \hline \end{array}$	r. $\begin{array}{r} 1 \\ 2 \\ \hline \end{array}$	s. $\begin{array}{r} 3 \\ 3 \\ \hline \end{array}$	t. $\begin{array}{r} 9 \\ 8 \\ \hline \end{array}$
u. $\begin{array}{r} 12 \\ 2 \\ \hline \end{array}$	v. $\begin{array}{r} 11 \\ 3 \\ \hline \end{array}$	w. $\begin{array}{r} 10 \\ 0 \\ \hline \end{array}$	x. $\begin{array}{r} 9 \\ 5 \\ \hline \end{array}$	y. $\begin{array}{r} 4 \\ 4 \\ \hline \end{array}$

Name _____

51

GA1392

Answer Sheet 3

a.	b.	c.	d.	e.
$\begin{array}{r} 5 \\ +\ 8 \\ \hline 13 \end{array}$	$\begin{array}{r} 7 \\ \times\ 7 \\ \hline 49 \end{array}$	$\begin{array}{r} 6 \\ -\ 3 \\ \hline 3 \end{array}$	$\begin{array}{r} 9 \\ \times\ 1 \\ \hline 9 \end{array}$	$\begin{array}{r} 4 \\ \times\ 10 \\ \hline 40 \end{array}$
f.	g.	h.	i.	j.
$\begin{array}{r} 3 \\ -\ 2 \\ \hline 1 \end{array}$	$\begin{array}{r} 10 \\ -\ 7 \\ \hline 3 \end{array}$	$\begin{array}{r} 8 \\ \times\ 8 \\ \hline 64 \end{array}$	$\begin{array}{r} 6 \\ +\ 7 \\ \hline 13 \end{array}$	$\begin{array}{r} 4 \\ +\ 9 \\ \hline 13 \end{array}$
k.	l.	m.	n.	o.
$\begin{array}{r} 8 \\ \times\ 10 \\ \hline 80 \end{array}$	$\begin{array}{r} 1 \\ \times\ 0 \\ \hline 0 \end{array}$	$\begin{array}{r} 10 \\ \times\ 10 \\ \hline 100 \end{array}$	$\begin{array}{r} 8 \\ \times\ 9 \\ \hline 72 \end{array}$	$\begin{array}{r} 7 \\ +\ 1 \\ \hline 8 \end{array}$
p.	q.	r.	s.	t.
$\begin{array}{r} 3 \\ +\ 5 \\ \hline 8 \end{array}$	$\begin{array}{r} 10 \\ \times\ 3 \\ \hline 30 \end{array}$	$\begin{array}{r} 1 \\ +\ 2 \\ \hline 3 \end{array}$	$\begin{array}{r} 3 \\ \times\ 3 \\ \hline 9 \end{array}$	$\begin{array}{r} 9 \\ +\ 8 \\ \hline 17 \end{array}$
u.	v.	w.	x.	y.
$\begin{array}{r} 12 \\ \times\ 2 \\ \hline 24 \end{array}$	$\begin{array}{r} 11 \\ -\ 3 \\ \hline 8 \end{array}$	$\begin{array}{r} 10 \\ \times\ 0 \\ \hline 0 \end{array}$	$\begin{array}{r} 9 \\ \times\ 5 \\ \hline 45 \end{array}$	$\begin{array}{r} 4 \\ +\ 4 \\ \hline 8 \end{array}$

Name _____

GA1392

Drill Sheet 4

a. 15 2	b. 10 3	c. 20 20	d. 18 5	e. 11 6
f. 14 7	g. 19 2	h. 10 10	i. 12 10	j. 16 5
k. 19 4	l. 20 5	m. 12 11	n. 17 2	o. 19 0
p. 20 10	q. 19 1	r. 17 5	s. 20 2	t. 14 14
u. 15 10	v. 14 3	w. 12 3	x. 16 2	y. 12 12

Name _____

53

GA1392

Answer Sheet 4

a.	b.	c.	d.	e.
$\begin{array}{r} 15 \\ \times\ 2 \\ \hline 30 \end{array}$	$\begin{array}{r} 10 \\ -\ 3 \\ \hline 7 \end{array}$	$\begin{array}{r} 20 \\ +\ 20 \\ \hline 40 \end{array}$	$\begin{array}{r} 18 \\ -\ 5 \\ \hline 13 \end{array}$	$\begin{array}{r} 11 \\ +\ 6 \\ \hline 17 \end{array}$
f.	g.	h.	i.	j.
$\begin{array}{r} 14 \\ -\ 7 \\ \hline 7 \end{array}$	$\begin{array}{r} 19 \\ +\ 2 \\ \hline 21 \end{array}$	$\begin{array}{r} 10 \\ \times\ 10 \\ \hline 100 \end{array}$	$\begin{array}{r} 12 \\ -\ 10 \\ \hline 2 \end{array}$	$\begin{array}{r} 16 \\ +\ 5 \\ \hline 21 \end{array}$
k.	l.	m.	n.	o.
$\begin{array}{r} 19 \\ -\ 4 \\ \hline 15 \end{array}$	$\begin{array}{r} 20 \\ \times\ 5 \\ \hline 100 \end{array}$	$\begin{array}{r} 12 \\ -\ 11 \\ \hline 1 \end{array}$	$\begin{array}{r} 17 \\ +\ 2 \\ \hline 19 \end{array}$	$\begin{array}{r} 19 \\ \times\ 0 \\ \hline 0 \end{array}$
p.	q.	r.	s.	t.
$\begin{array}{r} 20 \\ +\ 10 \\ \hline 30 \end{array}$	$\begin{array}{r} 19 \\ +\ 1 \\ \hline 20 \end{array}$	$\begin{array}{r} 17 \\ +\ 5 \\ \hline 22 \end{array}$	$\begin{array}{r} 20 \\ \times\ 2 \\ \hline 40 \end{array}$	$\begin{array}{r} 14 \\ -\ 14 \\ \hline 0 \end{array}$
u.	v.	w.	x.	y.
$\begin{array}{r} 15 \\ +\ 10 \\ \hline 25 \end{array}$	$\begin{array}{r} 14 \\ -\ 3 \\ \hline 11 \end{array}$	$\begin{array}{r} 12 \\ \times\ 3 \\ \hline 36 \end{array}$	$\begin{array}{r} 16 \\ \times\ 2 \\ \hline 32 \end{array}$	$\begin{array}{r} 12 \\ -\ 12 \\ \hline 0 \end{array}$

Name _____

54

Drill Sheet 5

a. 14 2	b. 18 0	c. 16 3	d. 15 3	e. 18 2
f. 17 2	g. 20 5	h. 16 0	i. 19 1	j. 15 4
k. 20 4	l. 15 5	m. 17 1	n. 14 4	o. 16 1
p. 20 20	q. 15 2	r. 18 4	s. 20 3	t. 14 0
u. 16 2	v. 18 3	w. 15 1	x. 15 15	y. 20 10

Name _____

GA1392

Answer Sheet 5

a. $\begin{array}{r} 14 \\ \times\ 2 \\ \hline 28 \end{array}$	**b.** $\begin{array}{r} 18 \\ \times\ 0 \\ \hline 0 \end{array}$	**c.** $\begin{array}{r} 16 \\ -\ 3 \\ \hline 13 \end{array}$	**d.** $\begin{array}{r} 15 \\ +\ 3 \\ \hline 18 \end{array}$	**e.** $\begin{array}{r} 18 \\ \times\ 2 \\ \hline 36 \end{array}$
f. $\begin{array}{r} 17 \\ -\ 2 \\ \hline 15 \end{array}$	**g.** $\begin{array}{r} 20 \\ \times\ 5 \\ \hline 100 \end{array}$	**h.** $\begin{array}{r} 16 \\ \times\ 0 \\ \hline 0 \end{array}$	**i.** $\begin{array}{r} 19 \\ \times\ 1 \\ \hline 19 \end{array}$	**j.** $\begin{array}{r} 15 \\ +\ 4 \\ \hline 19 \end{array}$
k. $\begin{array}{r} 20 \\ \times\ 4 \\ \hline 80 \end{array}$	**l.** $\begin{array}{r} 15 \\ +\ 5 \\ \hline 20 \end{array}$	**m.** $\begin{array}{r} 17 \\ -\ 1 \\ \hline 16 \end{array}$	**n.** $\begin{array}{r} 14 \\ -\ 4 \\ \hline 10 \end{array}$	**o.** $\begin{array}{r} 16 \\ \times\ 1 \\ \hline 16 \end{array}$
p. $\begin{array}{r} 20 \\ +\ 20 \\ \hline 40 \end{array}$	**q.** $\begin{array}{r} 15 \\ +\ 2 \\ \hline 17 \end{array}$	**r.** $\begin{array}{r} 18 \\ +\ 4 \\ \hline 22 \end{array}$	**s.** $\begin{array}{r} 20 \\ \times\ 3 \\ \hline 60 \end{array}$	**t.** $\begin{array}{r} 14 \\ -\ 0 \\ \hline 14 \end{array}$
u. $\begin{array}{r} 16 \\ +\ 2 \\ \hline 18 \end{array}$	**v.** $\begin{array}{r} 18 \\ +\ 3 \\ \hline 21 \end{array}$	**w.** $\begin{array}{r} 15 \\ \times\ 1 \\ \hline 15 \end{array}$	**x.** $\begin{array}{r} 15 \\ -\ 15 \\ \hline 0 \end{array}$	**y.** $\begin{array}{r} 20 \\ +\ 10 \\ \hline 30 \end{array}$

Name _____

Drill Sheet 6

a. 20 10	b. 24 11	c. 22 2	d. 24 2	e. 25 1
f. 22 4	g. 25 10	h. 20 8	i. 25 0	j. 23 3
k. 25 2	l. 22 5	m. 23 2	n. 24 3	o. 25 5
p. 25 11	q. 24 4	r. 23 12	s. 25 3	t. 20 0
u. 23 10	v. 25 4	w. 22 10	x. 20 1	y. 23 4

Name _____

57

GA1392

Answer Sheet 6

a. $\begin{array}{r} 20 \\ -\ 10 \\ \hline 10 \end{array}$	b. $\begin{array}{r} 24 \\ +\ 11 \\ \hline 35 \end{array}$	c. $\begin{array}{r} 22 \\ \times\ 2 \\ \hline 44 \end{array}$	d. $\begin{array}{r} 24 \\ \times\ 2 \\ \hline 48 \end{array}$	e. $\begin{array}{r} 25 \\ \times\ 1 \\ \hline 25 \end{array}$
f. $\begin{array}{r} 22 \\ +\ 4 \\ \hline 26 \end{array}$	g. $\begin{array}{r} 25 \\ -\ 10 \\ \hline 15 \end{array}$	h. $\begin{array}{r} 20 \\ +\ 8 \\ \hline 28 \end{array}$	i. $\begin{array}{r} 25 \\ +\ 0 \\ \hline 25 \end{array}$	j. $\begin{array}{r} 23 \\ +\ 3 \\ \hline 26 \end{array}$
k. $\begin{array}{r} 25 \\ \times\ 2 \\ \hline 50 \end{array}$	l. $\begin{array}{r} 22 \\ +\ 5 \\ \hline 27 \end{array}$	m. $\begin{array}{r} 23 \\ \times\ 2 \\ \hline 46 \end{array}$	n. $\begin{array}{r} 24 \\ +\ 3 \\ \hline 27 \end{array}$	o. $\begin{array}{r} 25 \\ -\ 5 \\ \hline 20 \end{array}$
p. $\begin{array}{r} 25 \\ +\ 11 \\ \hline 36 \end{array}$	q. $\begin{array}{r} 24 \\ +\ 4 \\ \hline 28 \end{array}$	r. $\begin{array}{r} 23 \\ +\ 12 \\ \hline 35 \end{array}$	s. $\begin{array}{r} 25 \\ -\ 3 \\ \hline 22 \end{array}$	t. $\begin{array}{r} 20 \\ \times\ 0 \\ \hline 0 \end{array}$
u. $\begin{array}{r} 23 \\ -\ 10 \\ \hline 13 \end{array}$	v. $\begin{array}{r} 25 \\ +\ 4 \\ \hline 100 \end{array}$	w. $\begin{array}{r} 22 \\ -\ 10 \\ \hline 12 \end{array}$	x. $\begin{array}{r} 20 \\ \times\ 1 \\ \hline 20 \end{array}$	y. $\begin{array}{r} 23 \\ +\ 4 \\ \hline 27 \end{array}$

Name _____

GA1392

2-Minute Drill

Each participant will have exactly 2 minutes or 120 seconds to successfully complete this four-step drill.

Step 1 contains 4 basic facts.

Step 2 contains 8 basic facts.

Step 3 contains 16 basic facts.

Step 4 contains 32 basic facts.

The drill is made up of addition, subtraction, multiplication, division and a combination of the four operations.

The teacher or drill leader will set a stopwatch at 00 seconds and then inform the participants that they will have 8 seconds to complete Step 1. After 8 seconds the leader will stop the watch, and the participants will put all pencils down on the leader's command: STOP! Any problem without an answer should be marked with an x.

The leader will inform the participants that the allotted time for Step 2 is 16 seconds. The STOP! command means all pencils down. As before, all problems without an answer are to be marked with an x.

This same procedure is to be followed for Steps 3 and 4.

Step 3 is allotted 32 seconds, and Step 4 is allotted 64 seconds.

The actual drill time is 2 minutes or 120 seconds.

GA1392

At the conclusion of the 2-minute period, each participant is to exchange his/her paper with a neighbor. The teacher or drill leader will review all sixty problems by giving the correct answer to each one. Instruct the class that all incorrect responses, as well as problems without answers, are to be marked with an x.

At this point, each child counts the number of correct answers.

57-60 correct .... super

53-56 correct .... great

48-52 correct .... good

44-47 correct .... fair

GA1392

Addition

Looks like a piece of cake to me!

Step 1: 8 seconds

3 + 9	8 + 7	6 + 5	7 + 3

Step 2: 16 seconds

10 + 0	4 + 9	7 + 7	6 + 10	2 + 9	8 + 9	9 + 3	11 + 1

Step 3: 32 seconds

12 + 0	9 + 9	3 + 5	4 + 3	8 + 4	10 + 10	2 + 11	10 + 2

5 + 9	7 + 8	9 + 0	1 + 9	10 + 2	3 + 6	7 + 4	2 + 8

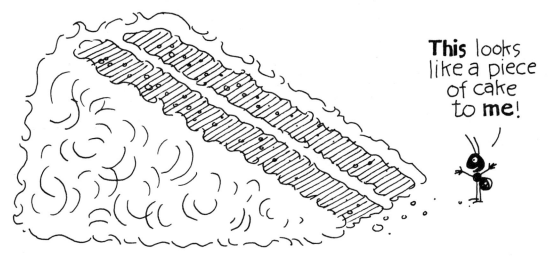

This looks like a piece of cake to **me!**

GA1392

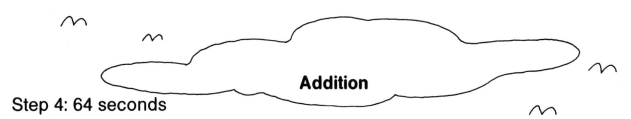

Addition

Step 4: 64 seconds

| 6
+ 7 | 4
+ 8 | 5
+ 0 | 7
+ 5 | 6
+ 0 | 4
+ 4 | 7
+ 9 | 10
+ 5 |

| 12
+ 3 | 11
+ 11 | 9
+ 8 | 3
+ 7 | 4
+ 10 | 6
+ 6 | 12
+ 2 | 9
+ 4 |

| 4
+ 12 | 10
+ 3 | 5
+ 5 | 11
+ 4 | 12
+ 12 | 5
+ 2 | 3
+ 10 | 11
+ 3 |

| 12
+ 1 | 8
+ 9 | 6
+ 4 | 5
+ 3 | 8
+ 8 | 3
+ 8 | 4
+ 11 | 4
+ 4 |

62

GA1392

Subtraction

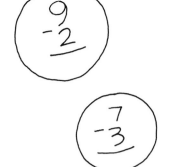

Step 1: 8 seconds

10 − 6	8 − 0	12 − 4	7 − 3

Step 2: 16 seconds

12 − 5	9 − 9	10 − 5	8 − 6	8 − 4	12 − 12	8 − 5	11 − 9

Step 3: 32 seconds

9 − 1	10 − 4	12 − 3	8 − 8	11 − 2	9 − 2	12 − 0	9 − 4

9 − 9	12 − 1	12 − 7	9 − 0	10 − 3	8 − 1	12 − 6	10 − 0

63

GA1392

Subtraction

Step 4: 64 seconds

10 - 5	8 - 6	12 - 1	9 - 9	6 - 3	8 - 3	11 - 5	7 - 6

9 - 1	6 - 1	12 - 0	7 - 4	11 - 8	8 - 1	10 - 3	9 - 0

10 - 4	7 - 7	11 - 6	6 - 5	9 - 8	12 - 4	5 - 3	9 - 6

11 - 7	7 - 5	12 - 3	6 - 4	8 - 0	10 - 1	9 - 7	12 - 11

12 flies minus 3 flies still makes a nice snack!

64

GA1392

Multiplication

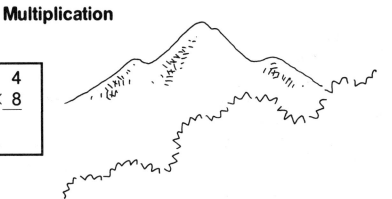

Step 1: 8 seconds

8 × 5	6 × 7	9 × 3	4 × 8

Step 2: 16 seconds

0 × 7	8 × 3	2 × 9	4 × 4	5 × 0	7 × 9	3 × 6	1 × 8

Step 3: 32 seconds

5 × 7	2 × 8	7 × 7	3 × 9	9 × 4	6 × 5	0 × 9	8 × 7

5 × 5	10 × 0	9 × 8	3 × 6	4 × 10	8 × 8	5 × 8	9 × 9

Whooooo knows
what 4×10 equals?

65

GA1392

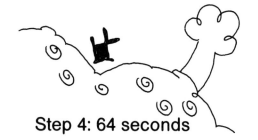

Multiplication

Step 4: 64 seconds

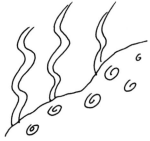

| 10
× 10 | 5
× 6 | 9
× 5 | 7
× 10 | 8
× 5 | 7
× 9 | 6
× 3 | 8
× 6 |

| 8
× 8 | 7
× 5 | 8
× 4 | 10
× 3 | 9
× 6 | 7
× 7 | 10
× 0 | 9
× 3 |

| 4
× 5 | 8
× 9 | 0
× 8 | 7
× 4 | 5
× 9 | 9
× 7 | 6
× 6 | 4
× 9 |

| 3
× 8 | 6
× 4 | 7
× 8 | 8
× 0 | 5
× 3 | 7
× 6 | 0
× 9 | 10
× 7 |

Five worms times three worms would tide me over 'til dinner!

66

GA1392

Division

Step 1: 8 seconds

8)16̄	5)20̄	8)8̄	6)36̄

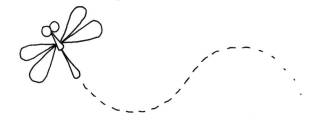

Step 2: 16 seconds

8)24̄	6)30̄	3)24̄	3)15̄	3)9̄	7)7̄	7)21̄	5)30̄

Step 3: 32 seconds

9)45̄	7)49̄	10)100̄	4)20̄	8)40̄	1)10̄	8)48̄	5)25̄

7)14̄	10)20̄	6)48̄	5)45̄	2)10̄	2)14̄	4)32̄	5)50̄

Nothin' like a side of **flies** with my lunch!

GA1392

Division

Step 4: 64 seconds

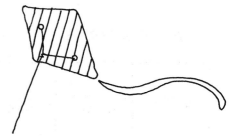

9⟌36	8⟌8	4⟌16	3⟌12	5⟌50	5⟌40	5⟌30	2⟌18

5⟌25	6⟌18	3⟌27	4⟌36	2⟌20	6⟌12	6⟌42	3⟌18

4⟌12	6⟌36	9⟌18	8⟌40	8⟌16	6⟌30	4⟌20	9⟌72

2⟌10	9⟌27	7⟌42	2⟌16	2⟌12	5⟌20	5⟌5	9⟌81

Guess I better pull the kite down and get crackin'...

68

GA1392

Addition, Subtraction, Multiplication, Division

Step 1: 8 seconds

8 × 2	7 − 3	5⟌15	9 + 7

Pencils ready...

Step 2: 16 seconds

4 × 8	12 − 8	5 × 8	6⟌36	4⟌40	9 × 0	10 − 10	10 × 0

Step 3: 32 seconds

7 × 3	10 + 2	8⟌32	10 − 8	8 × 8	3 + 9	9⟌63	6 × 5

4 × 3	6⟌24	4 + 10	9 × 5	10 − 4	8 × 7	11 − 3	12 − 11

Hang on! We're not ready yet!

69

GA1392

Addition, Subtraction, Multiplication, Division

| 6
× 7 | 12
− 3 | 9
+ 9 | 9⟌72 | 3
+ 10 | 9
× 8 | 10⟌100 | 7
× 8 |

| 3
+ 4 | 5⟌20 | 10
− 10 | 12
− 7 | 7⟌28 | 8
+ 8 | 10
× 0 | 11
− 0 |

| 10
+ 0 | 8
× 9 | 6⟌36 | 6
+ 9 | 7
× 6 | 11
− 2 | 2⟌10 | 12
− 5 |

| 0
× 10 | 10
− 4 | 8
× 7 | 11
− 9 | 6⟌42 | 8
+ 8 | 4⟌28 | 6
+ 7 |

64 seconds... on the nose!

70

GA1392

The Number Wheel . . . Part 1

The wheel illustrated below is divided into eight equal parts. Within each fractional portion of the wheel is both a letter and a number.

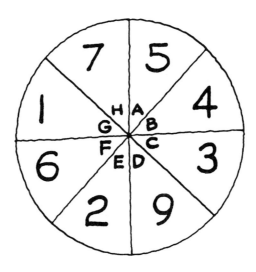

figure 1

The problems presented throughout this drill are shown through the use of various fractional representations of various parts of the wheel. These fractional parts will contain letters but will not show numbers.

Examples:

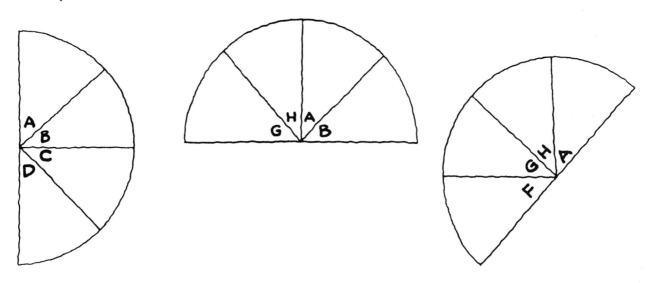

Perhaps the sum of each example is required. The student would then need to identify each letter with its missing number.

GA1392

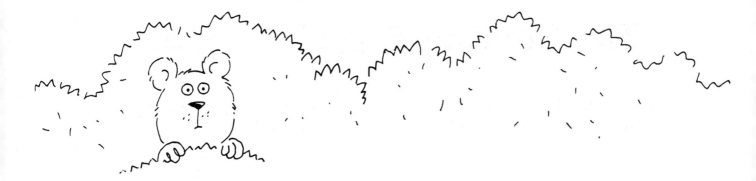

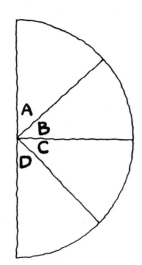

Step 1: Record each letter and its missing number.

A = 5

B = 4

C = 3

D = 9

Step 2: Find the sum of the four addends.

$5 + 4 + 3 + 9 = \boxed{21}$

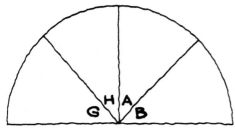

Step 1: Once again, refer to figure 1.

G =

H =

A =

B =

Step 2: Find the sum of the four addends.

___ + ___ + ___ + ___ = ☐

GA1392

Refer to figure 1.

Step 1: Record each letter and its missing number.

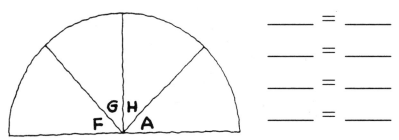

_____ = _____

_____ = _____

_____ = _____

_____ = _____

Step 2: Find the sum of the four addends.

___ + ___ + ___ + ___ = ☐

The above procedure in the examples should be followed when working with configurations that represent half the wheel.

73

GA1392

Work Sheet 1A

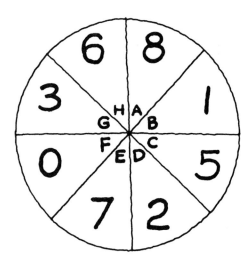

Name _____

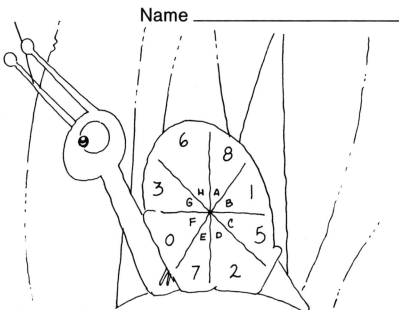

Step 1: Record each letter and its missing number.

_____ = _____

_____ = _____

_____ = _____

_____ = _____

Step 2: Find the sum of the four addends.

____ + ____ + ____ + ____ = ☐

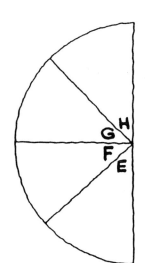

Step 1: Record each letter and its missing number.

_____ = _____

_____ = _____

_____ = _____

_____ = _____

Step 2: Find the sum of the four addends.

____ + ____ + ____ + ____ = ☐

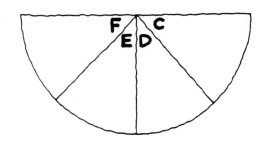

GA1392

Work Sheet 1B

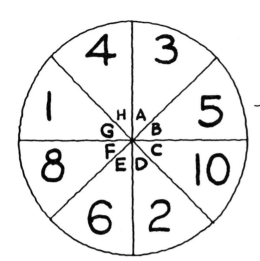

Step 1: Record each letter and its missing number.

_____ = _____

_____ = _____

_____ = _____

_____ = _____

Step 2: Find the sum of the four addends.

___ + ___ + ___ + ___ = ☐

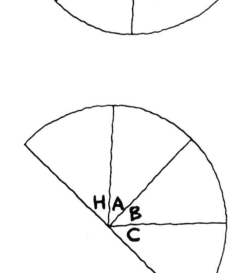

Step 1: Record each letter and its missing number.

_____ = _____

_____ = _____

_____ = _____

_____ = _____

Step 2: Find the sum of the four addends.

___ + ___ + ___ + ___ = ☐

75

GA1392

Work Sheet 1C

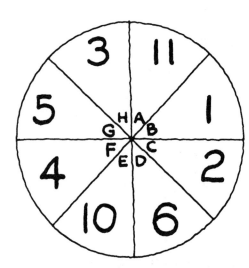

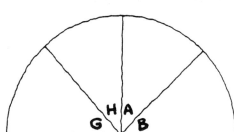

Step 1: Record each letter and its missing number.

_____ = _____

_____ = _____

_____ = _____

_____ = _____

Step 2: Find the sum of the four addends.

___ + ___ + ___ + ___ = ☐

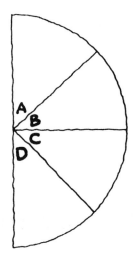

Step 1: Record each letter and its missing number.

_____ = _____

_____ = _____

_____ = _____

_____ = _____

Step 2: Find the sum of the four addends.

___ + ___ + ___ + ___ = ☐

GA1392

Work Sheet 1D

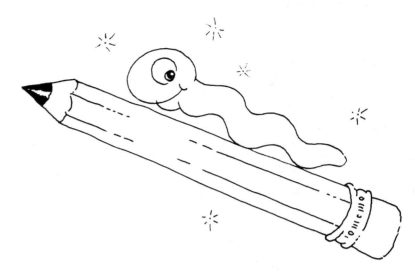

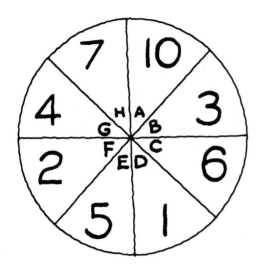

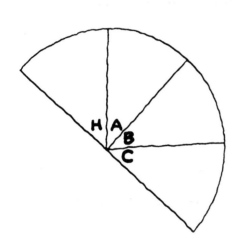

Step 1: Record each letter and its missing number.

_____ = _____

_____ = _____

_____ = _____

_____ = _____

Step 2: Find the sum of the four addends.

___ + ___ + ___ + ___ = ☐

Step 1: Record each letter and its missing number.

_____ = _____

_____ = _____

_____ = _____

_____ = _____

Step 2: Find the sum of the four addends.

___ + ___ + ___ + ___ = ☐

GA1392

Work Sheet 1E

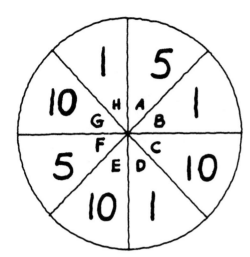

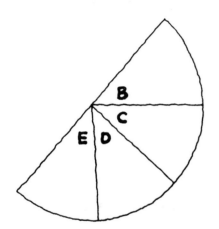

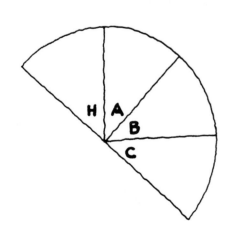

Name _____

STAY ON YOUR HALF!

Step 1: Record each letter and its missing number.

____ = ____

____ = ____

____ = ____

____ = ____

Step 2: Find the sum of the four addends.

___ + ___ + ___ + ___ = ☐

Step 1: Record each letter and its missing number.

____ = ____

____ = ____

____ = ____

____ = ____

Step 2: Find the sum of the four addends.

___ + ___ + ___ + ___ = ☐

GA1392

Use this work sheet and record addends of your choice.

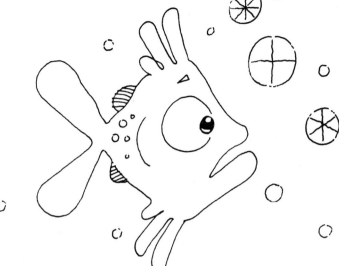

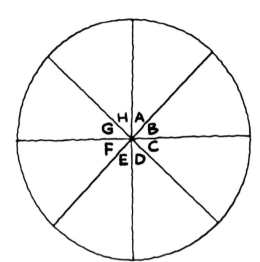

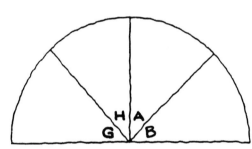

Step 1: Record each letter and its missing number.

_____ = _____

_____ = _____

_____ = _____

_____ = _____

Step 2: Find the sum of the four addends.

___ + ___ + ___ + ___ = ☐

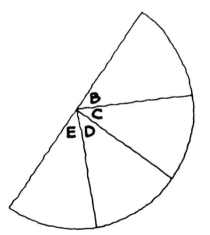

Step 1: Record each letter and its missing number.

_____ = _____

_____ = _____

_____ = _____

_____ = _____

Step 2: Find the sum of the four addends.

___ + ___ + ___ + ___ = ☐

GA1392

Use this work sheet and record addends of your choice.

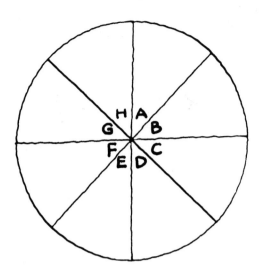

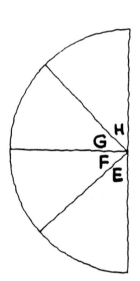

Step 1: Record each letter and its missing number.

_____ = _____

_____ = _____

_____ = _____

_____ = _____

Step 2: Find the sum of the four addends.

____ + ____ + ____ + ____ = ☐

Step 1: Record each letter and its missing number.

_____ = _____

_____ = _____

_____ = _____

_____ = _____

Step 2: Find the sum of the four addends.

____ + ____ + ____ + ____ = ☐

GA1392

The Number Wheel . . . Part 2

This section of the drill is based upon the idea of the wheel being divided into eight equal parts. Each separate part of the wheel should be referred to a part that is one-eighth of the total number wheel. Recorded within each fractional portion of the wheel is both a letter and a number.

The problems presented in The Number Wheel . . . Part 2 will not be confined to only the use of configurations representing half the number wheel. In Part 2, the student might see any of the following fractional configurations.

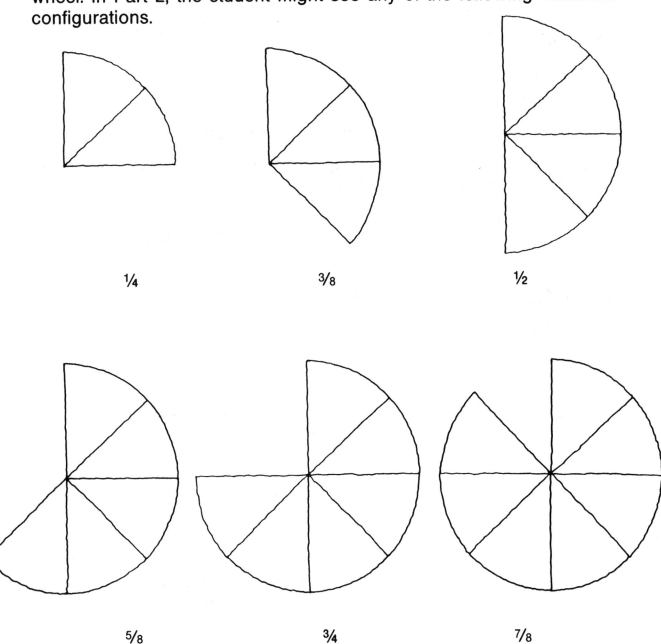

¼   ⅜   ½

⅝   ¾   ⅞

GA1392

Each of the configurations can be identified by letters. All of the sample problems can be solved by referring to the number wheel as shown in figure 2.

figure 2

Sample problems: Addition

Record the letter and its missing number. Then find the sum of the addends. Remember, use figure 2.

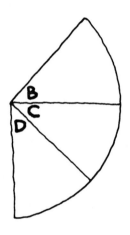

When recording the letters, you would write

B

C

D

Recording the missing numbers, you would write

B = 2

C = 6

D = 1

Find the sum of the three addends.

2 + 6 + 1 = 9

GA1392

Work Sheet 2A

Name _____

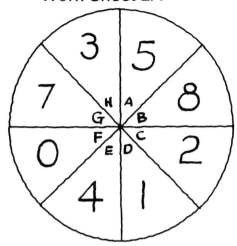

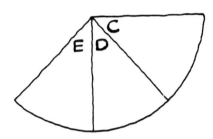

Step 1: Record each letter and its missing number.

_____ = _____

_____ = _____

_____ = _____

Step 2: Find the sum of the three addends.

____ + ____ + ____ = ☐

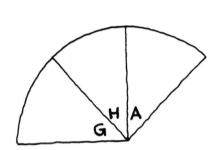

Step 1: Record each letter and its missing number.

_____ = _____

_____ = _____

_____ = _____

Step 2: Find the sum of the three addends.

____ + ____ + ____ = ☐

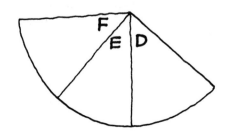

Step 1: Record each letter and its missing number.

_____ = _____

_____ = _____

_____ = _____

Step 2: Find the sum of the three addends.

____ + ____ + ____ = ☐

GA1392

Additional sample problems with reference to figure 2.

Here, doggies!

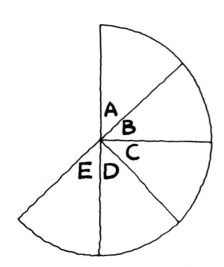

Step 1: Record each letter and its missing number.

_____ = _____   _____ = _____

_____ = _____   _____ = _____

_____ = _____

Step 2: Find the sum of the five addends.

_____ + _____ + _____ + _____ + _____ = ☐

Step 1: Record each letter and its missing number.

_____ = _____

_____ = _____

_____ = _____

Step 2: Find the sum of the three addends.

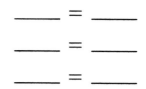

_____ + _____ + _____ = ☐

GA1392

Work Sheet 2B

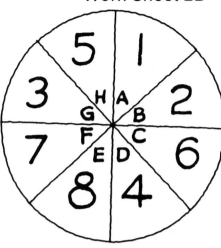

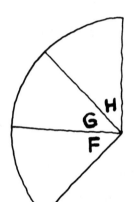

Step 1: Record each letter and its missing number.

____ = ____

____ = ____

____ = ____

Step 2: Find the sum of the three addends.

____ + ____ + ____ = ☐

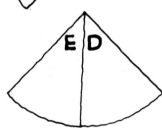

Step 1: Record each letter and its missing number.

____ = ____

____ = ____

Step 2: Find the sum of the two addends.

____ + ____ = ☐

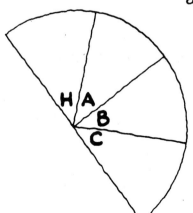

Step 1: Record each letter and its missing number.

____ = ____

____ = ____

____ = ____

____ = ____

Step 2: Find the sum of the four addends.

____ + ____ + ____ + ____

= ☐

GA1392

Work Sheet 2C

Name _____

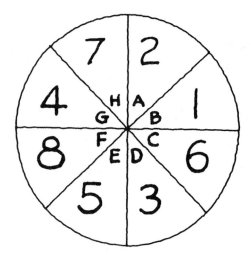

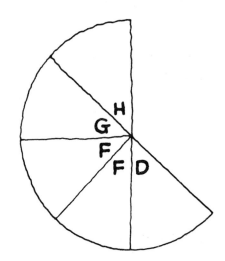

Step 1: Record each letter and its missing number.

____ = ____   ____ = ____

____ = ____   ____ = ____

____ = ____

Step 2: Find the sum of the five addends.

____ + ____ + ____ + ____ + ____ = □

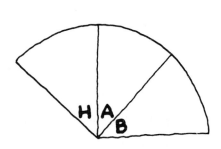

Step 1: Record each letter and its missing number.

____ = ____   ____ = ____   ____ = ____

Step 2: Find the sum of the three addends.

____ + ____ + ____ = □

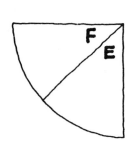

Step 1: Record each letter and its missing number.

____ = ____

____ = ____

Step 2: Find the sum of the two addends.

____ + ____ = □

GA1392

Work Sheet 2D

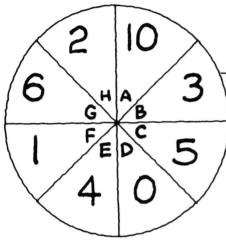

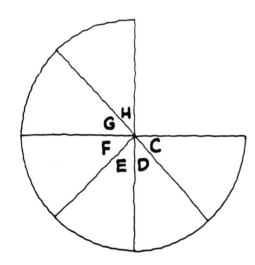

Step 1: Record each letter and its missing number.

_____ = _____   _____ = _____

_____ = _____   _____ = _____

_____ = _____   _____ = _____

Step 2: Find the sum of the six addends.

___ + ___ + ___ + ___ + ___ + ___ = □

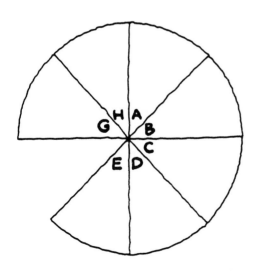

Step 1: Record each letter and its missing number.

_____ = _____   _____ = _____

_____ = _____   _____ = _____

_____ = _____   _____ = _____

_____ = _____

Step 2: Find the sum of the seven addends.

___ + ___ + ___ + ___ + ___ + ___ + ___

= □

GA1392

Work Sheet 2E

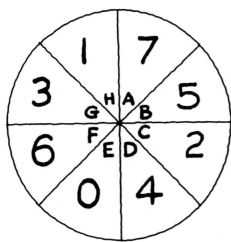

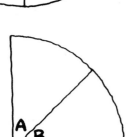

Step 1: Record each letter and its missing number.

____ = ____

____ = ____

Step 2: Find the sum of the two addends.

___ + ___ = ☐

Step 1: Record each letter and its missing number.

____ = ____

____ = ____

____ = ____

____ = ____

____ = ____

Step 2: Find the sum of the five addends.

___ + ___ + ___ + ___ + ___ = ☐

GA1392

Name _____

Use this work sheet and record addends of your choice.

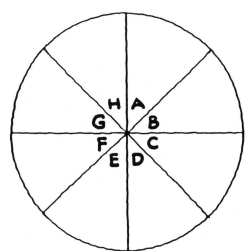

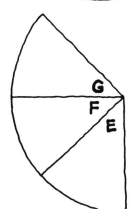

Step 1: Record each letter and its missing number.

____ = ____

____ = ____

____ = ____

Step 2: Find the sum of the three addends.

___ + ___ + ___ = ☐

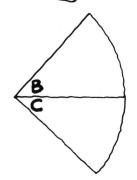

Step 1: Record each letter and its missing number.

____ = ____

____ = ____

Step 2: Find the sum of the two addends.

___ + ___ = ☐

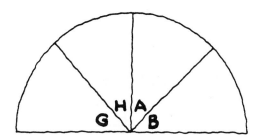

Step 1: Record each letter and its missing number.

____ = ____   ____ = ____

____ = ____   ____ = ____

Step 2: Find the sum of the four addends.

___ + ___ + ___ + ___ = ☐

89

GA1392

Use this work sheet and record addends of your choice.

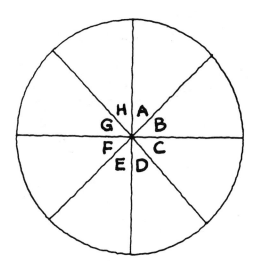

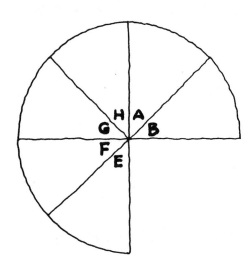

Step 1: Record each letter and its missing number.

_____ = _____

_____ = _____

_____ = _____

_____ = _____

_____ = _____

_____ = _____

Step 2: Find the sum of the six addends.

____ + ____ + ____ + ____ + ____ + ____ = ☐

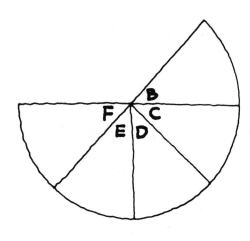

Step 1: Record each letter and its missing number.

_____ = _____

_____ = _____

_____ = _____

_____ = _____

_____ = _____

Step 2: Find the sum of the five addends.

____ + ____ + ____ + ____ + ____ = ☐

90

The Number Wheel . . . Part 3

Each fractional portion is divided into eight equal parts. Each portion has a letter and a number. Refer to the fractional configurations and . . .

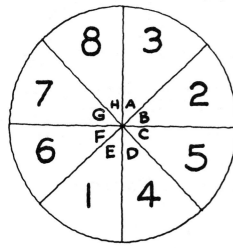

Step 1: Record each letter and its missing number.

Step 2: Find the sum of the addends.

Step 3: Subtract the sum from a constant number— 29.

Example: Refer to figure 3.

figure 3

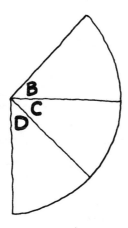

Step 1:

$$\underline{B} = \underline{2}$$
$$\underline{C} = \underline{5}$$
$$\underline{D} = \underline{4}$$

Step 2:

$$\underline{2} + \underline{5} + \underline{4} = 11$$

Step 3:

$$\begin{array}{r} 29 \\ -\ 11 \\ \hline \boxed{18} \end{array}\ \text{answer}$$

GA1392

Work Sheet 3A

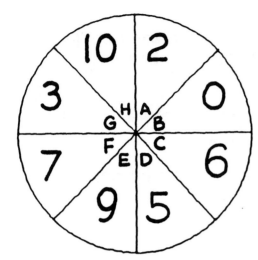

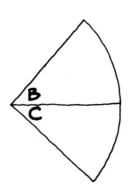

Step 1: Record.

_____ = _____

_____ = _____

Step 2: Find sum.

___ + ___ = ☐

Step 3: Subtract from 29.

29

- ☐

answer

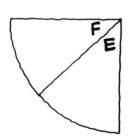

Step 1: Record.

_____ = _____

_____ = _____

Step 2: Find sum.

___ + ___ = ☐

Step 3: Subtract from 29.

29

- ☐

answer

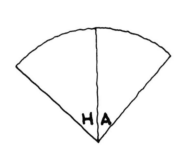

Step 1: Record.

_____ = _____

_____ = _____

Step 2: Find sum.

___ + ___ = ☐

Step 3: Subtract from 29.

29

- ☐

answer

GA1392

Work Sheet 3B

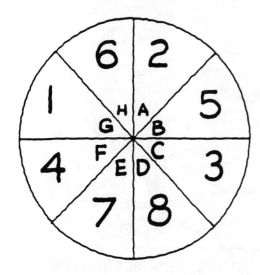

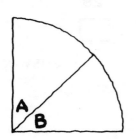

Step 1: Record.

____ = ____

____ = ____

Step 2: Find sum.

____ + ____ = ☐

Step 3: Subtract from 29.

29
- ☐
_____

answer

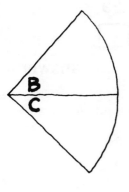

Step 1: Record.

____ = ____

____ = ____

Step 2: Find sum.

____ + ____ = ☐

Step 3: Subtract from 29.

29
- ☐
_____

answer

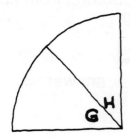

Step 1: Record.

____ = ____

____ = ____

Step 2: Find sum.

____ + ____ = ☐

Step 3: Subtract from 29.

29
- ☐
_____

answer

GA1392

Work Sheet 3C

Name _____

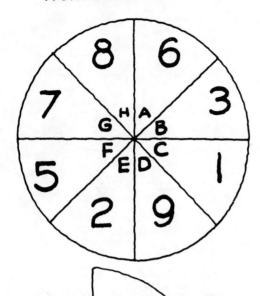

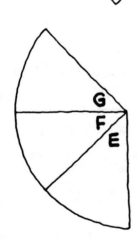

Step 1: Record.

____ = ____

____ = ____

____ = ____

Step 2: Find sum.

____ + ____ + ____ = ☐

Step 3: Subtract from 29.

29
-☐

answer

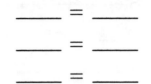

Step 1: Record.

____ = ____

____ = ____

____ = ____

Step 2: Find sum.

____ + ____ + ____ = ☐

Step 3: Subtract from 29.

29
-☐

answer

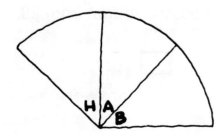

Step 1: Record.

____ = ____

____ = ____

____ = ____

Step 2: Find sum.

____ + ____ + ____ = ☐

Step 3: Subtract from 29.

29
-☐

answer

Name _____

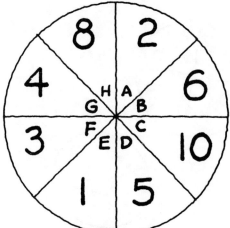

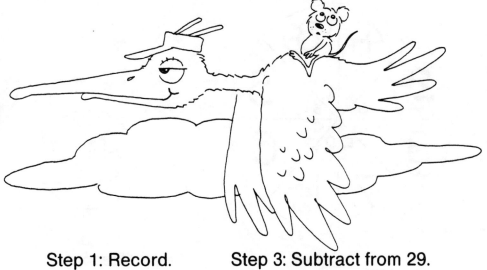

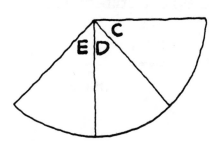

Step 1: Record.

_____ = _____

_____ = _____

_____ = _____

Step 2: Find sum.

____ + ____ + ____ = □

Step 3: Subtract from 29.

29
- □
_____
answer

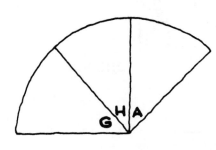

Step 1: Record.

_____ = _____

_____ = _____

_____ = _____

Step 2: Find sum.

____ + ____ + ____ = □

Step 3: Subtract from 29.

29
- □
_____
answer

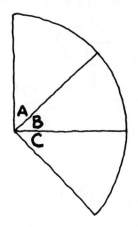

Step 1: Record.

_____ = _____

_____ = _____

_____ = _____

Step 2: Find sum.

____ + ____ + ____ = □

Step 3: Subtract from 29.

29
- □
_____
answer

GA1392

Work Sheet 3E

Name _____

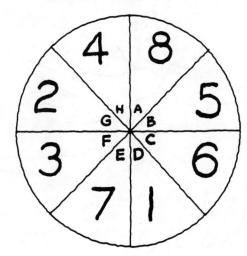

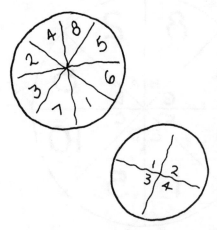

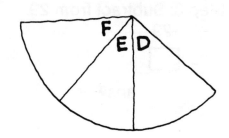

Step 1: Record.

_____ = _____

_____ = _____

_____ = _____

Step 2: Find sum.

____ + ____ + ____ = ▢

Step 3: Subtract from 29.

29

- ▢
_____

answer

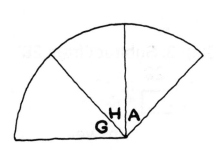

Step 1: Record.

_____ = _____

_____ = _____

_____ = _____

Step 2: Find sum.

____ + ____ + ____ = ▢

Step 3: Subtract from 29.

29

- ▢
_____

answer

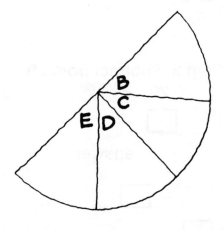

Step 1: Record.

_____ = _____

_____ = _____

_____ = _____

_____ = _____

Step 2: Find sum.

____ + ____ + ____ + ____ = ▢

Step 3: Subtract from 29.

29

- ▢
_____

answer

GA1392

Work Sheet 3F

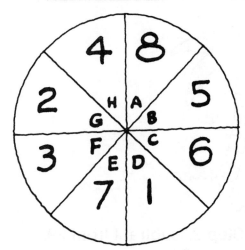

Name _____

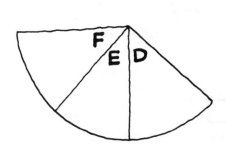

Step 1: Record.

_____ = _____

_____ = _____

_____ = _____

Step 2: Find sum.

___ + ___ + ___ = ☐

Step 3: Subtract from 29.

29

− ☐

answer

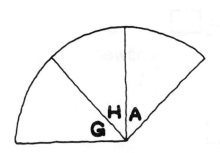

Step 1: Record.

_____ = _____

_____ = _____

_____ = _____

Step 2: Find sum.

___ + ___ + ___ = ☐

Step 3: Subtract from 29.

29

− ☐

answer

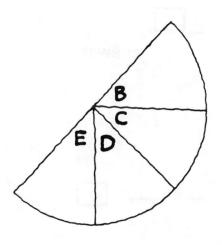

Step 1: Record.

_____ = _____

_____ = _____

_____ = _____

_____ = _____

Step 2: Find sum.

___ + ___ + ___ + ___ = ☐

Step 3: Subtract from 29.

29

− ☐

answer

97

GA1392

Work Sheet 3G

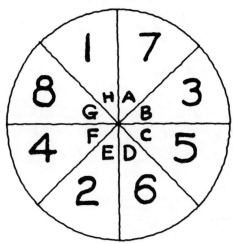

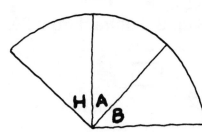

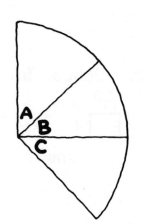

Step 1: Record.

____ = ____

____ = ____

____ = ____

Step 2: Find sum.

___ + ___ + ___ = ☐

Step 3: Subtract from 29.

29
- ☐
_____

answer

Step 1: Record.

____ = ____

____ = ____

____ = ____

Step 2: Find sum.

___ + ___ + ___ = ☐

Step 3: Subtract from 29.

29
- ☐
_____

answer

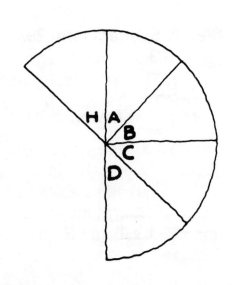

Step 1: Record.

____ = ____

____ = ____

____ = ____

____ = ____

____ = ____

Step 2: Find sum.

___ + ___ + ___ + ___ + ___ = ☐

Step 3: Subtract from 29.

29
- ☐
_____

answer

98

GA1392

Colors Count

Colors Count Box

4 brown	**1** green	**7** blue
8 red	**6** yellow	**3** purple
2 orange	**9** white	**5** pink

figure 1

The box above figure 1 is comprised of numbers and colors. Each number in the Colors Count Box is assigned a particular color.

When you refer to color, think about and refer to its number.

$$PURPLE = 3$$
$$PINK = 5$$
$$BROWN = 4$$

Throughout this drill, colors will combine with numbers to create basic computational skill facts.

Example: Refer to figure 1.

YELLOW
+ ORANGE

When making number substitutions for colors,

$$YELLOW = 6$$
$$+ ORANGE = 2$$

therefore, $6 + 2 = 8$

Additional examples:

$$YELLOW = 6$$
$$- ORANGE = 2$$
$$4$$

$$YELLOW = 6$$
$$\times ORANGE = 2$$
$$12$$

$$YELLOW \div ORANGE$$
$$6 \div 2 = 3$$

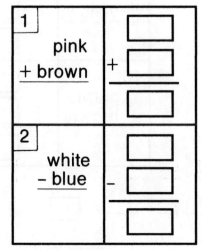

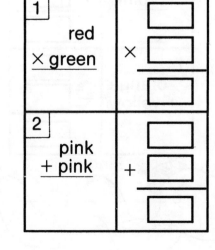

figure 2

The problems on the reproducible drill pages will look like those in figure 2.

When the drill pages of Colors Count are reproduced, have the student color in the numbers in the Colors Count Box.

GA1392

Colors Count

Addition facts using numbers 1 through 9—two addends

Colors Count Box

4 brown	1 green	7 blue
8 red	6 yellow	3 purple
2 orange	9 white	5 pink

1 brown + purple   +

2 orange + red   +

3 blue + orange   +

4 yellow + green   +

5 purple + orange   +

6 yellow + purple   +

7 green + red   +

8 blue + green   +

9 orange + orange   +

10 white + green   +

11 brown + brown   +

12 purple + pink   +

13 pink + pink   +

14 orange + yellow   +

15 brown + yellow   +

16 purple + brown   +

GA1392

Colors Count

Addition facts using numbers to 18—two addends

Colors Count Box

10 brown	4 green	3 blue
2 red	12 yellow	5 purple
16 orange	6 white	15 pink

1
yellow
+ blue

2
green
+ yellow

3
brown
+ purple

4
green
+ white

5
orange
+ red

6
pink
+ red

7
yellow
+ purple

8
white
+ red

9
brown
+ green

10
blue
+ pink

11
white
+ white

12
purple
+ white

13
yellow
+ white

14
brown
+ brown

15
yellow
+ red

16
pink
+ purple

COLORS
COUNT

Colors Count

Colors Count Box

Addition facts using numbers to 30—two addends

5 brown	**8** green	**20** blue
2 red	**10** yellow	**3** purple
15 orange	**25** white	**4** pink

1
yellow
+ brown
+ ☐ ☐ ☐

2
yellow
+ yellow
+ ☐ ☐ ☐

3
white
+ brown
+ ☐ ☐ ☐

4
orange
+ pink
+ ☐ ☐ ☐

5
blue
+ pink
+ ☐ ☐ ☐

6
red
+ white
+ ☐ ☐ ☐

7
blue
+ red
+ ☐ ☐ ☐

8
yellow
+ orange
+ ☐ ☐ ☐

9
yellow
+ blue
+ ☐ ☐ ☐

10
pink
+ blue
+ ☐ ☐ ☐

11
purple
+ blue
+ ☐ ☐ ☐

12
blue
+ green
+ ☐ ☐ ☐

13
white
+ pink
+ ☐ ☐ ☐

14
green
+ pink
+ ☐ ☐ ☐

15
orange
+ brown
+ ☐ ☐ ☐

16
green
+ yellow
+ ☐ ☐ ☐

GA1392

Colors Count

Addition facts to 18—three addends

Colors Count Box

4 brown	7 green	2 blue
8 red	1 yellow	6 purple
3 orange	5 white	9 pink

1 purple
   brown
 + brown

2 yellow
   brown
 + green

3 brown
   red
 + yellow

4 green
   red
 + yellow

5 pink
   yellow
 + red

6 purple
   purple
 + purple

7 green
   purple
 + orange

8 yellow
   purple
 + pink

9 yellow
   white
 + purple

10 blue
    pink
 + green

11 yellow
    yellow
 + blue

12 green
    purple
 + blue

13 blue
    green
 + pink

14 blue
    green
 + blue

15 pink
    yellow
 + blue

16 red
    red
 + yellow

GA1392

Name _____

Colors Count

Addition facts to 30—three addends

Colors Count Box

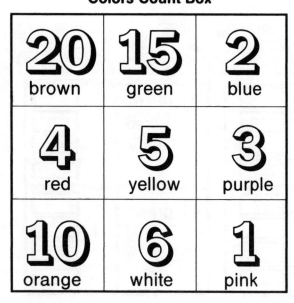

20 brown	**15** green	**2** blue
4 red	**5** yellow	**3** purple
10 orange	**6** white	**1** pink

1	purple orange + white	+
2	white white + yellow	+

3	orange orange + orange	+
4	brown red + white	+

5	orange white + pink	+
6	green yellow + orange	+

7	red blue + brown	+
8	orange red + pink	+

9	orange blue + pink	+
10	green yellow + pink	+

11	red red + white	+
12	red yellow + purple	+

13	red green + pink	+
14	orange pink + pink	+

15	orange pink + green	+
16	brown pink + purple	+

GA1392

Colors Count

Subtraction facts to 12

Colors Count Box

9 brown	**12** green	**5** blue
1 red	**10** yellow	**3** purple
11 orange	**0** white	**2** pink

1
yellow
- blue

2
brown
- brown

3
orange
- brown

4
brown
- purple

5
green
- purple

6
orange
- blue

7
yellow
- pink

8
green
- brown

9
green
red

10
orange
- purple

11
green
- white

12
brown
- blue

13
yellow
- red

14
green
- pink

15
yellow
- purple

16
brown
- pink

105

GA1392

Colors Count

Name _____

Subtraction facts to 18

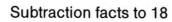

Colors Count Box

3 brown	**18** green	**14** blue
16 red	**2** yellow	**5** purple
1 orange	**15** white	**12** pink

1	green − brown	−
2	red − yellow	−

3	red − brown	−
4	white − purple	−

5	green − purple	−
6	white − pink	−

7	red − purple	−
8	green − orange	−

9	white − yellow	−
10	pink −orange	−

11	pink − brown	−
12	pink − yellow	−

13	green − yellow	−
14	pink − pink	−

15	red −orange	−
16	white − brown	−

GA1392

Colors Count

Subtraction facts to 24

Colors Count Box

0 brown	5 green	19 blue
21 red	24 yellow	3 purple
2 orange	18 white	1 pink

1
white
− pink

2
red
− red

3
yellow
− pink

4
blue
− purple

5
red
− pink

6
yellow
− red

7
yellow
− orange

8
white
− brown

9
yellow
− brown

10
white
−orange

11
white
−purple

12
blue
−orange

13
yellow
− purple

14
green
− pink

15
blue
− white

16
white
− white

GA1392

Colors Count

Name _____

Multiplication facts to 30

Colors Count Box

6 brown	**1** green	**4** blue
2 red	**7** yellow	**0** purple
5 orange	**3** white	**8** pink

1
orange
× red

× ▢
▢
———
▢

2
pink
× red

× ▢
▢
———
▢

3
brown
× green

× ▢
▢
———
▢

4
purple
× orange

× ▢
▢
———
▢

5
pink
× white

× ▢
▢
———
▢

6
blue
× blue

× ▢
▢
———
▢

7
blue
× brown

× ▢
▢
———
▢

8
white
× orange

× ▢
▢
———
▢

9
green
× blue

× ▢
▢
———
▢

10
green
× orange

× ▢
▢
———
▢

11
orange
× orange

× ▢
▢
———
▢

12
red
× brown

× ▢
▢
———
▢

13
white
× yellow

× ▢
▢
———
▢

14
yellow
× red

× ▢
▢
———
▢

15
brown
× orange

× ▢
▢
———
▢

16
yellow
× purple

× ▢
▢
———
▢

GA1392

Colors Count

Multiplication facts to 40

Colors Count Box

7 brown	10 green	8 blue
5 red	9 yellow	2 purple
6 orange	3 white	0 pink

1
green
× white

2
pink
× blue

3
orange
× orange

4
blue
× red

5
orange
× blue

6
red
× red

7
green
× purple

8
yellow
× red

9
pink
× green

10
blue
× orange

11
yellow
× white

12
purple
× yellow

13
brown
× brown

14
brown
× red

15
yellow
× pink

16
red
× yellow

Colors Count

Multiplication facts to 50

Colors Count Box

10 brown	**2** green	**9** blue
8 red	**4** yellow	**3** purple
6 orange	**5** white	**12** pink

1 blue × white

2 blue × yellow

3 red × orange

4 green × white

5 brown × purple

6 white × white

7 green blue

8 pink × green

9 red × yellow

10 brown × yellow

11 yellow ×orange

12 white × blue

13 yellow red

14 purple blue

15 white × brown

16 orange × purple

GA1392

Colors Count

Multiplication facts to 100

Colors Count Box

2 brown	**9** green	**3** blue
10 red	**4** yellow	**5** purple
6 orange	**8** white	**7** pink

1
red
× pink

×

2
red
× green

×

3
purple
× red

×

4
white
× white

×

5
pink
× white

×

6
brown
× pink

×

7
blue
× white

×

8
purple
× purple

×

9
yellow
× red

×

10
orange
× red

×

11
green
× yellow

×

12
orange
× white

×

13
green
× green

×

14
purple
× orange

×

15
pink
× yellow

×

16
red
× red

×

GA1392

Colors Count

Colors Count Box

Multiplication facts to 100

2 brown	**9** green	**3** blue
10 red	**4** yellow	**5** purple
6 orange	**8** white	**7** pink

17 blue × green ×

18 pink × pink ×

19 green × white ×

20 orange × green ×

21 red × yellow ×

22 white × pink ×

23 white × orange ×

24 green × red ×

25 pink × brown ×

26 orange × orange ×

27 green × orange ×

28 red × purple ×

29 green × blue ×

30 white × red ×

31 purple × pink ×

32 yellow × yellow ×

112

Colors Count

Addition, subtraction, multiplication—facts to 100

Colors Count Box

12 brown	**2** green	**6** blue
0 red	**3** yellow	**8** purple
18 orange	**4** white	**5** pink

1
purple
× yellow   ×

2
brown
– red   –

3
brown
– blue   –

4
orange
– white   –

5
brown
+ purple   +

6
yellow
× brown   ×

7
orange
+ red   +

8
blue
× blue   ×

9
blue
× pink   ×

10
purple
– pink   –

11
orange
– orange   –

12
brown
– pink   –

13
brown
× green   ×

14
orange
– pink   –

15
pink
× purple   ×

16
yellow
× yellow   ×

GA1392

Name _____

Colors Count

Addition, subtraction, multiplication—facts to 100

Colors Count Box

12 brown	**2** green	**6** blue
0 red	**3** yellow	**8** purple
18 orange	**4** white	**5** pink

17 orange − red − ☐ ☐ / ☐

18 white × pink × ☐ ☐ / ☐

19 orange − blue − ☐ ☐ / ☐

20 orange + purple + ☐ ☐ / ☐

21 orange − yellow − ☐ ☐ / ☐

22 red × brown × ☐ ☐ / ☐

23 purple × purple × ☐ ☐ / ☐

24 purple × red × ☐ ☐ / ☐

25 purple + blue + ☐ ☐ / ☐

26 green × brown × ☐ ☐ / ☐

27 yellow × purple × ☐ ☐ / ☐

28 orange + white + ☐ ☐ / ☐

29 orange + brown + ☐ ☐ / ☐

30 pink − red − ☐ ☐ / ☐

31 brown + yellow + ☐ ☐ / ☐

32 brown × red × ☐ ☐ / ☐

GA1392

Colors Count

Solving equations

Colors Count Box

2 brown	**5** green	**1** blue
8 red	**10** yellow	**4** purple
6 orange	**3** white	**7** pink

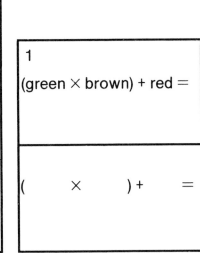

1

(green × brown) + red =

(×   ) +   =

2

(pink × white) + purple =

(×   ) +   =

3

(yellow × yellow) + red =

(×   ) +   =

4

(yellow × blue) − red =

(×   ) −   =

5

(orange × white) + pink =

(×   ) +   =

6

(pink × pink) + blue =

(×   ) +   =

7

(orange × brown) + yellow =

(×   ) +   =

GA1392

Tri-It

Tri-It makes use of the triangle. There is special emphasis on the sides of the triangle. Every triangle is made up of three line segments called sides. Throughout this drill, the sides will be labeled LEFT, RIGHT and BOTTOM.

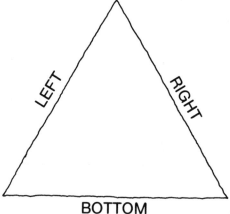

Tri-It is a two-step drill that requires finding the sum of the numbers on each side of the triangle.

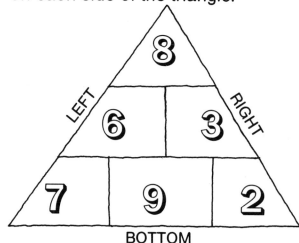

LEFT = 8 + 6 + 7 = 21

RIGHT = 8 + 3 + 2 = 13

BOTTOM = 7 + 9 + 2 = 18

After the sum of each side is determined, record this information in the Data Box. Record both the side and its sum in a rank order from high to low. Use L for LEFT, R for RIGHT and B for BOTTOM.

GA1392

Data Box

Side	Sum
L	21
B	18
R	13

Example:

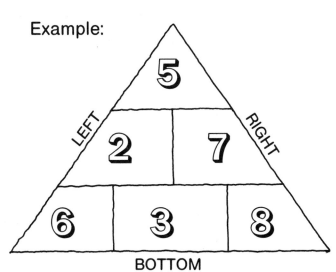

LEFT

RIGHT

BOTTOM

LEFT = 6 + 2 + 5 = 13

RIGHT = 5 + 7 + 8 = 20

BOTTOM = 6 + 3 + 8 = 17

Data Box

Side	Sum
R	20
B	17
L	13

GA1392

Tri-It Work Sheet

a.

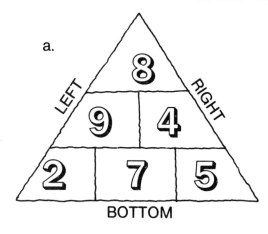

LEFT =

RIGHT =

BOTTOM =

Data Box

Side	Sum

b.

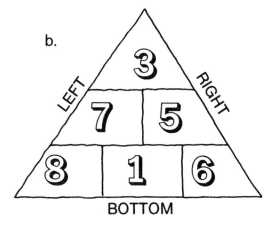

LEFT =

RIGHT =

BOTTOM =

Data Box

Side	Sum

c.

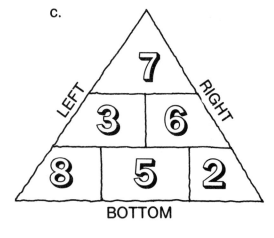

LEFT =

RIGHT =

BOTTOM =

Data Box

Side	Sum

GA1392

Tri-It Work Sheet

d.

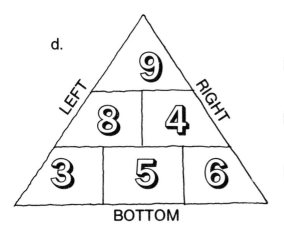

LEFT =

RIGHT =

BOTTOM =

Data Box

Side	Sum

e.

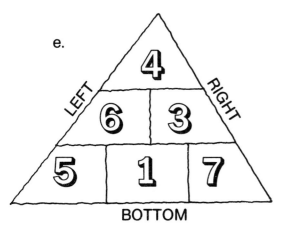

LEFT =

RIGHT =

BOTTOM =

Data Box

Side	Sum

f.

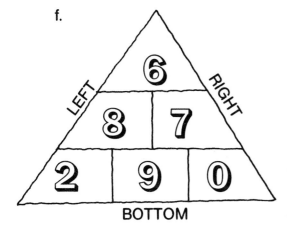

LEFT =

RIGHT =

BOTTOM =

Data Box

Side	Sum

GA1392

Tri-It Work Sheet

g.

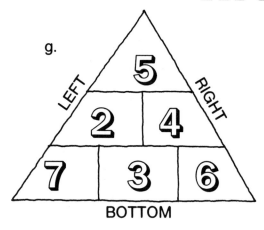

LEFT =

RIGHT =

BOTTOM =

Data Box

Side	Sum

h.

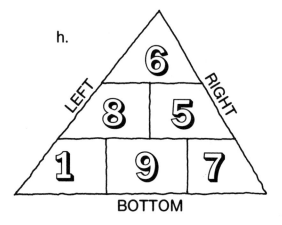

LEFT =

RIGHT =

BOTTOM =

Data Box

Side	Sum

i.

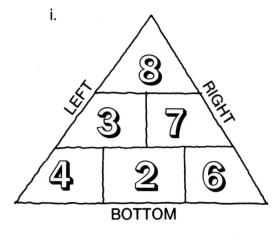

LEFT =

RIGHT =

BOTTOM =

Data Box

Side	Sum

GA1392

Tri-It Work Sheet

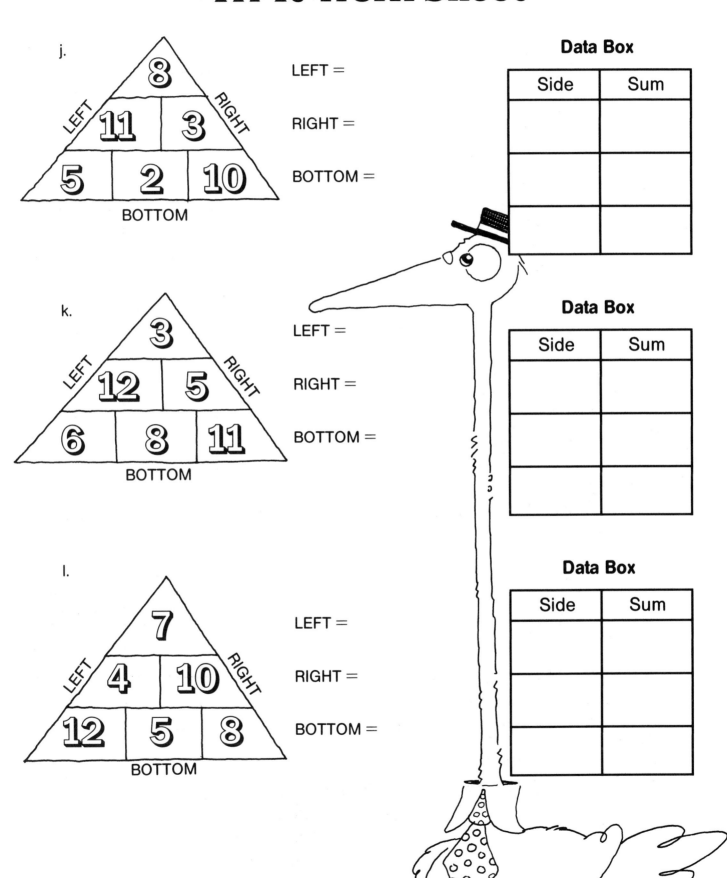

j.

LEFT =

RIGHT =

BOTTOM =

Data Box

Side	Sum

k.

LEFT =

RIGHT =

BOTTOM =

Data Box

Side	Sum

l.

LEFT =

RIGHT =

BOTTOM =

Data Box

Side	Sum

121

GA1392

Tri-It Work Sheet

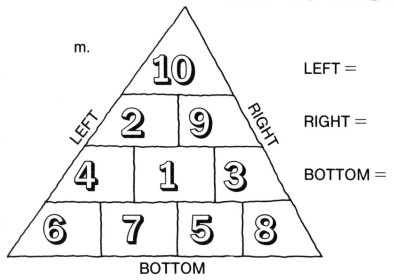

m.

LEFT =

RIGHT =

BOTTOM =

Data Box

Side	Sum

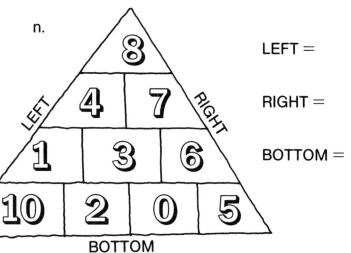

n.

LEFT =

RIGHT =

BOTTOM =

Data Box

Side	Sum

o.

LEFT =

RIGHT =

BOTTOM =

Data Box

Side	Sum

122

GA1392

Tri-It Work Sheet

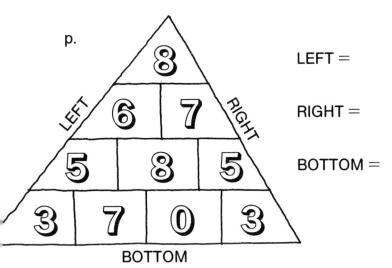

p.

LEFT =

RIGHT =

BOTTOM =

Data Box

Side	Sum

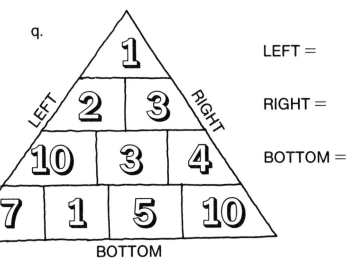

q.

LEFT =

RIGHT =

BOTTOM =

Data Box

Side	Sum

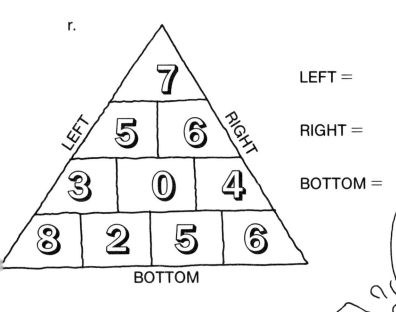

r.

LEFT =

RIGHT =

BOTTOM =

Data Box

Side	Sum

GA1392

Create your own Tri-It Work Sheet.

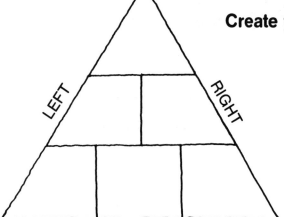

LEFT =

RIGHT =

BOTTOM =

Data Box

Side	Sum

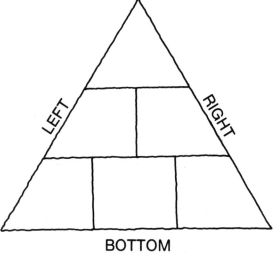

LEFT =

RIGHT =

BOTTOM =

Data Box

Side	Sum

LEFT =

RIGHT =

BOTTOM =

Data Box

Side	Sum

GA1392

Self-Serv

The Self-Serv is comprised of a rectangle with four numbers. There is one number in each corner of the drill rectangle. Within the drill rectangle is space for five different problems. These problems are to be created by the student. The student can create and solve the five problems in the drill rectangle. An operation sign is also included. This sign will be used by the student to arrive at answers to the problems. The student will create the problems by recording any two of the numbers that can be found in the four corners. If the operation word is *addition*, then the recorded numbers in the individual problem spaces will become addends. The student will then find the sum of each problem. If the operation word is *multiplication*, then the recorded numbers will become factors, and the student will find the product of each problem. All pages in this drill will contain four rectangles and twenty problems.

Example:

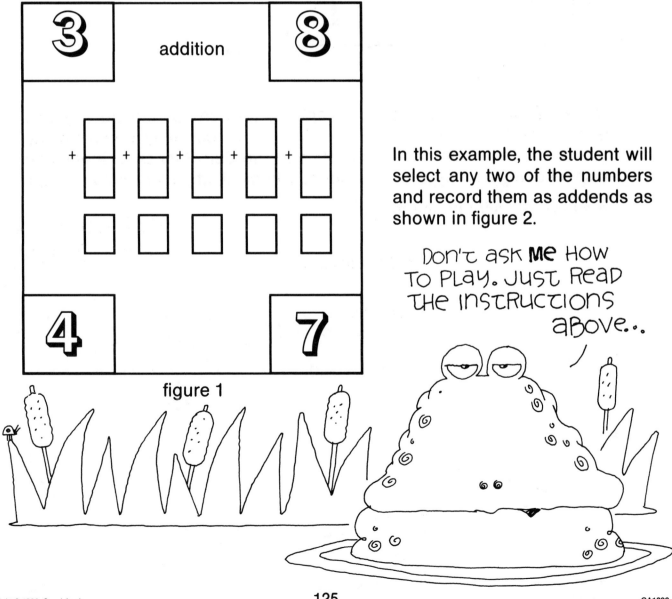

figure 1

In this example, the student will select any two of the numbers and record them as addends as shown in figure 2.

Don't ask ME How To PLaY. JUST ReaD THe InSTRUCTIONS aBoVe...

125

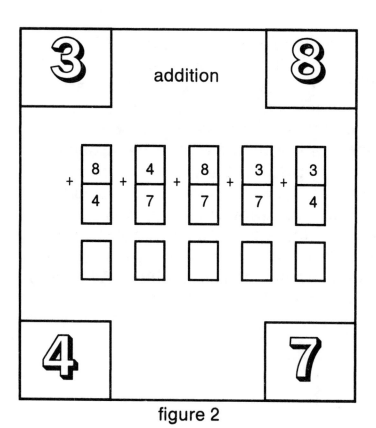

addition

figure 2

Figure 2 shows the addends as selected by the student.
The student will now work on finding the sums of the problems he/she created.

Once the twenty problems are completed (or perhaps the student was given a two or three-minute time limit to successfully complete the problems), the student should exchange the Self-Serv Drill Sheet with a sheet completed by another student. Check all of the correct answers.

A suggested scoring might be:

17-20 Just Super

15-16 Good

13-14 Not Bad

Self-Serv... why didn't I think of that?

GA1392

Self-Serv Drill Sheet

Choose any two corner numbers and record them as addends in each of the five problem spaces.

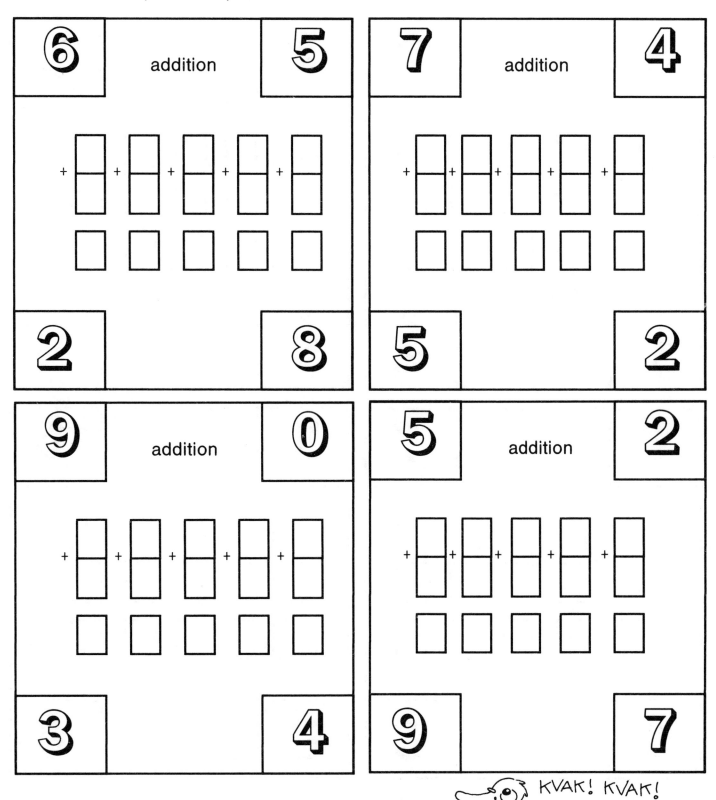

Name _____

Self-Serv Drill Sheet

Choose any two corner numbers and record them as addends in each of the five problem spaces.

9	addition	3

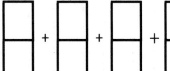

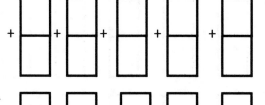

$+ \square\ +\ \square\ +\ \square\ +\ \square\ +\ \square$

1		2

8	addition	2

6		3

5	addition	7

$+ \square\ +\ \square\ +\ \square\ +\ \square\ +\ \square$

1	addition	3

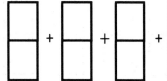
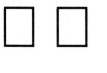
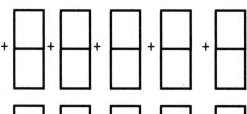

4		8

8		6

GA1392

Self-Serv Drill Sheet

Choose any two corner numbers and record them as addends in each of the five problem spaces.

4   addition   **2**

+ ☐   + ☐   + ☐   + ☐   + ☐

☐   ☐   ☐   ☐   ☐

9   **8**

0   addition   **8**

+ ☐   + ☐   + ☐   + ☐   + ☐

☐   ☐   ☐   ☐   ☐

4   **7**

7   addition   **3**

+ ☐   + ☐   + ☐   + ☐   + ☐

☐   ☐   ☐   ☐   ☐

2   **5**

2   addition   **6**

+ ☐   + ☐   + ☐   + ☐   + ☐

☐   ☐   ☐   ☐   ☐

9   **8**

GA1392

Name _____

Self-Serv Drill Sheet

Choose any two corner numbers and record them as addends in each of the five problem spaces.

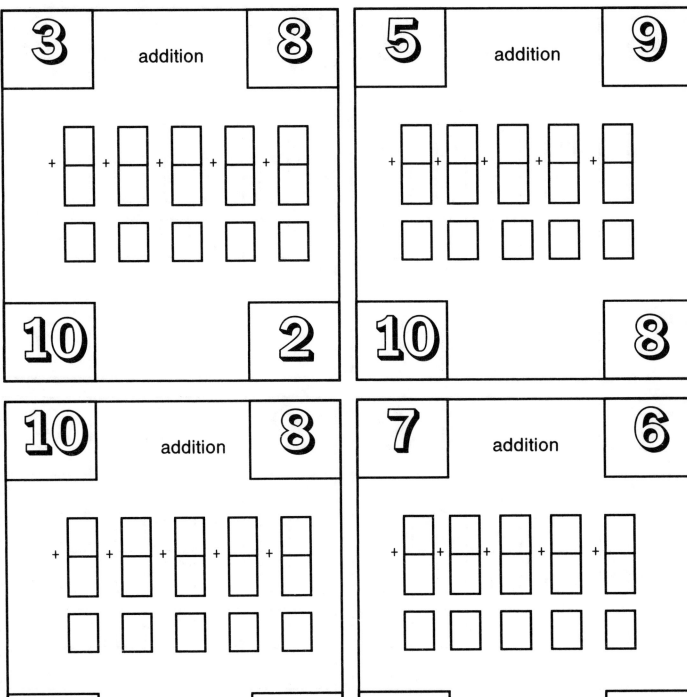

3   addition   **8**

10   **2**

5   addition   **9**

10   **8**

10   addition   **8**

4   **5**

7   addition   **6**

10   **5**

GA1392

Name _____

Self-Serv Drill Sheet

Write your own numbers in each of the four corners and record them as addends in each of the five problem spaces.

addition

$+\square$ $+\square$ $+\square$ $+\square$ $+\square$

addition

$+\square$ $+\square$ $+\square$ $+\square$ $+\square$

addition

$+\square$ $+\square$ $+\square$ $+\square$ $+\square$

addition

$+\square$ $+\square$ $+\square$ $+\square$ $+\square$

GA1392

Self-Serv Drill Sheet

Choose any two corner numbers and record them as factors in each of the five problem spaces.

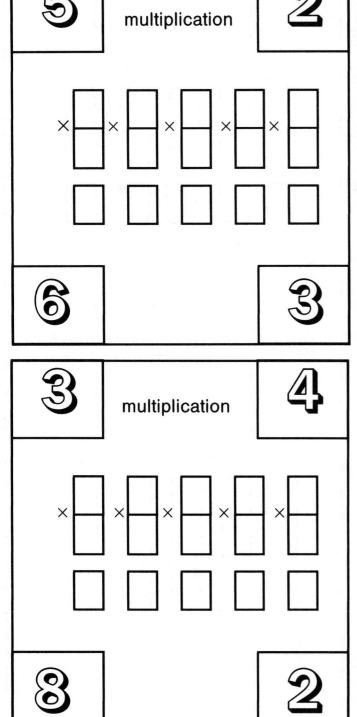

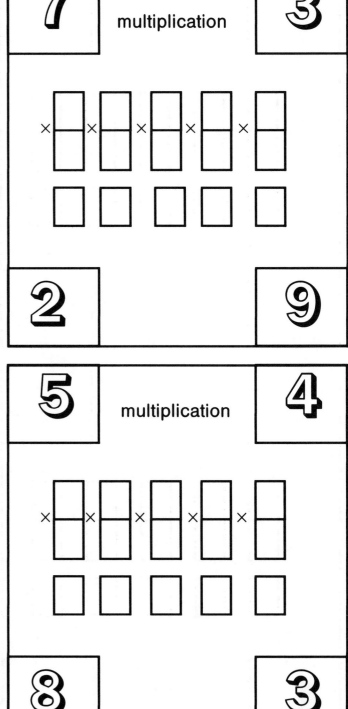

132

Name _____

Self-Serv Drill Sheet

Choose any two corner numbers and record them as factors in each
of the five problem spaces.

 4   multiplication   **2**

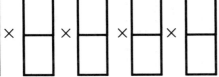

3   **5**

6   multiplication   **4**

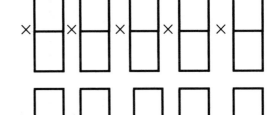

1   **2**

5   multiplication   **3**

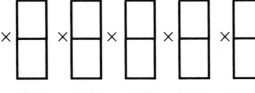

1   **6**

3   multiplication   **5**

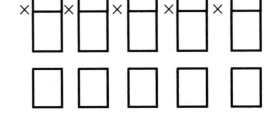

6   **4**

GA1392

Self-Serv Drill Sheet

Choose any two corner numbers and record them as factors in each of the five problem spaces.

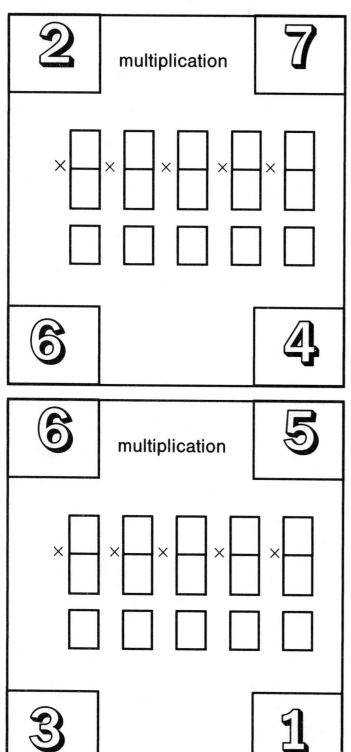

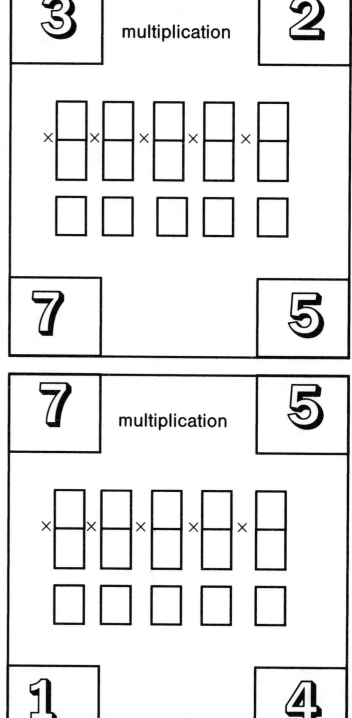

Self-Serv Drill Sheet

Choose any two corner numbers and record them as factors in each of the five problem spaces.

5 multiplication **3**

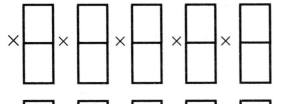

$\times \square$ $\times \square$ $\times \square$ $\times \square$ $\times \square$

9 **2**

9 multiplication **4**

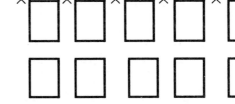

$\times \square$ $\times \square$ $\times \square$ $\times \square$ $\times \square$

6 **3**

7 multiplication **9**

$\times \square$ $\times \square$ $\times \square$ $\times \square$ $\times \square$

2 **4**

7 multiplication **9**

$\times \square$ $\times \square$ $\times \square$ $\times \square$ $\times \square$

5 **8**

GA1392

Self-Serv Drill Sheet

Choose any two corner numbers and record them as factors in each of the five problem spaces.

multiplication

× □ × □ × □ × □ × □

□ □ □ □ □

multiplication

× □ × □ × □ × □ × □

□ □ □ □ □

multiplication

× □ × □ × □ × □ × □

□ □ □ □ □

multiplication

× □ × □ × □ × □ × □

□ □ □ □ □

GA1392

D.D.T. (Daily Drill Test Taking)

This drill will test both your knowledge of as well as your ability to successfully complete problems in basic math facts. In this section are drill sheets that contain ten rows of problems. Each page has one problem on row 1, two problems on row 2, etc. Each page contains fifty-five problems. The allotted time for completion of all the basic facts on each page is three minutes.

The teacher, using a stopwatch, will give the word *begin.* You must now try to correctly solve the problems on the drill page beginning with the problem on row 1. As you complete a row, move on to the next row. You will have three minutes to try to successfully complete the ten rows. When the teacher gives the command *stop*, put your pencil down.

Before the teacher begins reading correct answers to the problems, you give your paper to your neighbor to check, and take your neighbor's paper so that you may check it. Your job is to listen carefully and circle all incorrect answers found on your neighbor's page.

SCORING

If you were working on the tenth row when the teacher gave the command *stop*, that's super!

If you were working on the ninth row, that's great!

If you were working on the eighth row, that's good!

Even more important is the number of correct answers.

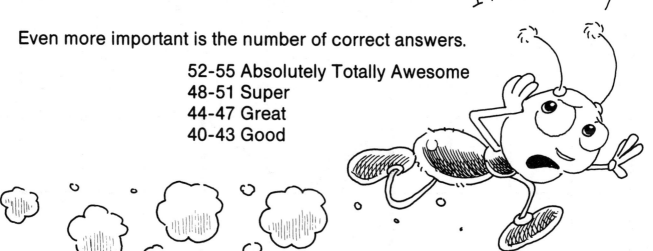

52-55 Absolutely Totally Awesome
48-51 Super
44-47 Great
40-43 Good

GA1392

Name _____

1	9 +3 ☐								
2	2 +10 ☐	7 +5 ☐							
3	11 +1 ☐	4 +9 ☐	3 +7 ☐						
4	1 +9 ☐	10 +10 ☐	3 +4 ☐	4 +7 ☐					
5	6 +8 ☐	8 +2 ☐	7 +8 ☐	5 +8 ☐	8 +3 ☐				
6	8 +5 ☐	6 +6 ☐	8 +0 ☐	11 +11 ☐	11 +2 ☐	5 +7 ☐			
7	10 +0 ☐	8 +6 ☐	6 +5 ☐	8 +7 ☐	7 +7 ☐	9 +6 ☐	9 +4 ☐		
8	4 +3 ☐	9 +0 ☐	2 +8 ☐	12 +0 ☐	5 +6 ☐	9 +5 ☐	0 +6 ☐	7 +6 ☐	
9	3 +9 ☐	7 +4 ☐	8 +8 ☐	4 +5 ☐	2 +11 ☐	6 +0 ☐	10 +2 ☐	6 +9 ☐	4 +8 ☐
10	0 +8 ☐	5 +4 ☐	0 +12 ☐	3 +8 ☐	9 +9 ☐	0 +8 ☐	4 +9 ☐	2 +10 ☐	5 +9 ☐	6 +7 ☐

Copyright © 1992, Good Apple

138

GA1392

#										
1	11 − 5 □									
2	8 − 0 □	9 − 9 □								
3	7 − 6 □	12 − 2 □	10 − 5 □							
4	9 − 7 □	7 − 1 □	11 − 3 □	10 − 10 □						
5	7 − 5 □	5 − 4 □	9 − 8 □	6 − 5 □	9 − 5 □					
6	7 − 7 □	9 − 1 □	5 − 3 □	10 − 9 □	5 − 5 □	8 − 4 □				
7	6 − 6 □	10 − 1 □	8 − 1 □	12 − 1 □	9 − 4 □	11 − 2 □	8 − 6 □			
8	11 − 1 □	8 − 5 □	9 − 2 □	7 − 7 □	8 − 3 □	6 − 3 □	10 − 3 □	6 − 2 □		
9	8 − 2 □	9 − 9 □	10 − 4 □	8 − 7 □	7 − 4 □	11 − 0 □	9 − 6 □	10 − 2 □	6 − 1 □	
10	9 − 3 □	12 − 0 □	8 − 8 □	7 − 3 □	10 − 6 □	7 − 2 □	12 − 9 □	6 − 4 □	5 − 2 □	12 − 8 □

GA1392

#									
1	7×5								
2	8×3	4×5							
3	5×5	6×7	9×3						
4	7×7	8×2	6×3	7×1					
5	3×5	4×4	6×5	8×1	10×4				
6	4×9	3×7	2×10	10×10	10×9	6×6			
7	9×8	7×9	1×8	3×9	9×5	6×9	4×8		
8	3×0	2×2	5×8	9×9	5×2	1×10	0×4	8×6	
9	8×9	0×8	2×9	2×8	1×4	10×3	8×5	8×10	4×2
10	3×2	3×10	2×7	4×7	6×8		1×9	8×4	8×8	10×10

140

Name _____

1	$2\overline{)10}$									
2	$9\overline{)18}$	$3\overline{)12}$								
3	$5\overline{)50}$	$2\overline{)16}$	$2\overline{)20}$							
4	$5\overline{)40}$	$8\overline{)32}$	$2\overline{)30}$	$3\overline{)18}$						
5	$9\overline{)45}$	$6\overline{)24}$	$7\overline{)35}$	$5\overline{)10}$	$3\overline{)21}$					
6	$7\overline{)42}$	$10\overline{)20}$	$4\overline{)12}$	$6\overline{)30}$	$10\overline{)50}$	$5\overline{)25}$				
7	$4\overline{)16}$	$1\overline{)10}$	$2\overline{)14}$	$6\overline{)18}$	$8\overline{)40}$	$4\overline{)44}$	$3\overline{)15}$			
8	$7\overline{)21}$	$5\overline{)35}$	$8\overline{)16}$	$5\overline{)45}$	$10\overline{)10}$	$3\overline{)9}$	$4\overline{)40}$			
9	$5\overline{)30}$	$6\overline{)12}$	$4\overline{)24}$	$3\overline{)30}$	$2\overline{)24}$	$4\overline{)8}$	$4\overline{)32}$	$3\overline{)6}$	$2\overline{)8}$	
10	$10\overline{)40}$	$2\overline{)18}$	$2\overline{)40}$	$2\overline{)12}$	$6\overline{)42}$	$7\overline{)14}$	$2\overline{)6}$	$8\overline{)8}$	$6\overline{)6}$	$5\overline{)15}$

Copyright © 1992, Good Apple

141

GA1392

1
- $5\overline{)20} = \square$

2
- $12 - 8 = \square$
- $3\overline{)15} = \square$

3
- $11 - 0 = \square$
- $7 \times 9 = \square$
- $8 \times 3 = \square$

4
- $10 \times 4 = \square$
- $11 - 3 = \square$
- $8 \times 7 = \square$
- $15 - 7 = \square$

5
- $8\overline{)24} = \square$
- $2\overline{)20} = \square$
- $8 + 8 = \square$
- $4 \times 8 = \square$
- $11 \times 1 = \square$

6
- $4\overline{)16} = \square$
- $1 \times 10 = \square$
- $2\overline{)16} = \square$
- $9 - 9 = \square$
- $8 + 7 = \square$
- $4\overline{)28} = \square$

7
- $3 \times 10 = \square$
- $7 \times 8 = \square$
- $5 \times 5 = \square$
- $2 \times 10 = \square$
- $3 \times 8 = \square$
- $6 \times 6 = \square$
- $12 - 6 = \square$

8
- $12 - 12 = \square$
- $6\overline{)6} = \square$
- $10 \times 10 = \square$
- $6\overline{)24} = \square$
- $4\overline{)20} = \square$
- $10 \times 3 = \square$
- $5\overline{)15} = \square$
- $7\overline{)28} = \square$

9
- $10\overline{)30} = \square$
- $15 - 6 = \square$
- $7 + 8 = \square$
- $8\overline{)16} = \square$
- $4 + 4 = \square$
- $9 \times 7 = \square$
- $12 - 5 = \square$
- $7 \times 7 = \square$
- $7\overline{)49} = \square$

10
- $8 \times 4 = \square$
- $10\overline{)20} = \square$
- $3\overline{)30} = \square$
- $0 \times 10 = \square$
- $4\overline{)8} = \square$
- $8\overline{)40} = \square$
- $6 \times 10 = \square$
- $5 - 5 = \square$
- $5\overline{)40} = \square$
- $11 - 7 = \square$

GA1392

This page is for you to create your own daily drill.

GA1392

#										
1	12 + 0 = □									
2	4 + 8 = □	9 + 9 = □								
3	4 + 9 = □	7 + 6 = □	7 + 8 = □							
4	9 + 6 = □	7 + 3 = □	11 + 1 = □	6 + 4 = □						
5	3 + 8 = □	8 + 6 = □	3 + 9 = □	8 + 1 = □	9 + 8 = □					
6	2 + 9 = □	9 + 7 = □	7 + 2 = □	4 + 7 = □	8 + 4 = □	6 + 6 = □				
7	9 + 1 = □	6 + 8 = □	8 + 7 = □	9 + 5 = □	7 + 4 = □	2 + 8 = □	6 + 9 = □			
8	6 + 7 = □	4 + 6 = □	0 + 8 = □	6 + 5 = □	5 + 5 = □	1 + 9 = □	5 + 8 = □	7 + 9 = □		
9	8 + 5 = □	0 + 12 = □	3 + 7 = □	8 + 8 = □	0 + 7 = □	10 + 10 = □	8 + 2 = □	5 + 7 = □	9 + 4 = □	
10	9 + 2 = □	7 + 7 = □	8 + 9 = □	7 + 5 = □	5 + 9 = □	8 + 3 = □	0 + 9 = □	0 + 6 = □	5 + 6 = □	9 + 3 = □

144

GA1392

1	9 − 8 □										
2	6 − 6 □	8 − 4 □									
3	9 − 4 □	10 − 9 □	8 − 1 □								
4	6 − 5 □	11 − 8 □	5 − 4 □	12 − 2 □							
5	10 − 5 □	11 − 4 □	12 − 3 □	11 − 3 □	10 − 7 □						
6	9 − 3 □	12 − 1 □	6 − 1 □	8 − 5 □	11 − 1 □	9 − 6 □					
7	8 − 0 □	8 − 2 □	9 − 7 □	12 − 0 □	12 − 6 □	10 − 1 □	4 − 4 □				
8	5 − 3 □	10 − 6 □	11 − 7 □	7 − 2 □	9 − 5 □	10 − 10 □	7 − 6 □	9 − 0 □			
9	11 − 0 □	5 − 5 □	8 − 3 □	11 − 9 □	10 − 8 □	7 − 7 □	10 − 2 □	11 − 5 □	10 − 0 □		
10	9 − 9 □	12 − 4 □	10 − 4 □	12 − 5 □	9 − 2 □	11 − 6 □	10 − 3 □	4 − 0 □	9 − 1 □	11 − 11 □	

145

GA1392

Name _____

1	7 ×3 ☐											
2	9 ×5 ☐	0 ×5 ☐										
3	8 ×6 ☐	2 ×9 ☐	8 ×8 ☐									
4	1 ×1 ☐	9 ×7 ☐	5 ×9 ☐	2 ×7 ☐								
5	4 ×0 ☐	4 ×6 ☐	3 ×3 ☐	10 ×0 ☐	1 ×1 ☐							
6	3 ×9 ☐	7 ×7 ☐	9 ×8 ☐	6 ×8 ☐	7 ×10 ☐	3 ×7 ☐						
7	10 ×9 ☐	6 ×3 ☐	7 ×8 ☐	4 ×3 ☐	9 ×3 ☐	6 ×9 ☐	9 ×9 ☐					
8	4 ×4 ☐	8 ×5 ☐	9 ×0 ☐	6 ×4 ☐	10 ×5 ☐	6 ×6 ☐	0 ×7 ☐	7 ×2 ☐				
9	8 ×9 ☐	10 ×8 ☐	3 ×4 ☐	7 ×9 ☐	2 ×2 ☐	4 ×9 ☐	10 ×10 ☐	8 ×7 ☐	9 ×1 ☐			
10	9 ×6 ☐	5 ×5 ☐	5 ×8 ☐	8 ×0 ☐	9 ×4 ☐	10 ×6 ☐	3 ×2 ☐	0 ×1 ☐	6 ×1 ☐	9 ×2 ☐		

146

GA1392

Name _____

#										
1	2⟌14									
2	6⟌12	7⟌42								
3	2⟌18	5⟌35	3⟌15							
4	10⟌30	2⟌10	6⟌18	5⟌25						
5	5⟌50	3⟌21	6⟌30	4⟌8	6⟌36					
6	2⟌16	10⟌10	3⟌9	4⟌16	1⟌7	4⟌20				
7	9⟌27	5⟌30	4⟌24	6⟌42	9⟌18	3⟌27	2⟌30			
8	5⟌15	10⟌50	5⟌30	2⟌12	7⟌35	3⟌12	5⟌20	1⟌10		
9	9⟌9	2⟌20	5⟌10	7⟌21	8⟌16	9⟌36	6⟌24	4⟌28	10⟌100	
10	2⟌24	4⟌36	7⟌28	7⟌14	10⟌20	4⟌12	4⟌32	3⟌18	8⟌32	6⟌30

GA1392

Name _____

#										
1	6 × 6 = □									
2	7 × 5 = □	12 − 3 = □								
3	9 × 4 = □	3 × 4 = □	10 − 7 = □							
4	15 − 8 = □	6 + 6 = □	7 + 10 = □	□ ÷ 7)42						
5	7 + 9 = □	14 − 7 = □	12 − 5 = □	14 − 6 = □	□ 7)42					
6	□ 5)20	3 + 8 = □	5 × 5 = □	8 + 4 = □	3 + 9 = □	5 × 7 = □				
7	10 − 6 = □	15 − 7 = □	6 × 8 = □	14 − 5 = □	□ 10)100	9 + 4 = □	10 × 0 = □			
8	□ 9)36	12 − 9 = □	4 + 9 = □	6 × 6 = □	□ 8)8	□ 4)20	10 × 10 = □	1 × 1 = □		
9	9 × 8 = □	4 + 9 = □	4 × 5 = □	8 + 9 = □	12 − 6 = □	□ 8)40	15 − 6 = □	7 × 7 = □	7 + 7 = □	
10	□ 6)42	10 + 5 = □	4 + 8 = □	15 − 5 = □	9 − 9 = □	0 × 9 = □	8 × 6 = □	8 × 8 = □	5 × 6 = □	7 + 8 = □

148

GA1392

Name _____

This page is for you to create your own daily drill.

Name _____

1	9 +9 □								
2	12 +3 □	10 +7 □							
3	9 +4 □	11 +3 □	13 +7 □						
4	5 +10 □	5 +7 □	10 +0 □	11 +1 □					
5	4 +9 □	12 +4 □	15 +1 □	10 +8 □	9 +3 □				
6	8 +5 □	6 +6 □	10 +6 □	13 +2 □	3 +12 □				
7	10 +1 □	12 +7 □	15 +0 □	10 +2 □	7 +5 □	8 +6 □	14 +1 □		
8	13 +5 □	5 +9 □	10 +9 □	5 +8 □	11 +2 □	12 +6 □	7 +4 □	10 +10 □	
9	2 +10 □	11 +0 □	4 +12 □	11 +4 □	12 +8 □	7 +6 □	8 +8 □	10 +3 □	4 +10 □
10	6 +7 □	12 +5 □	6 +8 □	10 +5 □	7 +7 □	3 +9 □	10 +4 □	4 +7 □	11 +9 □	9 +5 □

GA1392

1	12 − 1 ☐								
2	11 − 0 ☐	10 − 3 ☐							
3	8 − 8 ☐	9 − 9 ☐	12 − 7 ☐						
4	12 − 11 ☐	10 − 2 ☐	9 − 3 ☐						
5	8 − 7 ☐	9 − 4 ☐	12 − 2 ☐	10 − 6 ☐					
6	11 − 7 ☐	10 − 10 ☐	7 − 6 ☐	11 − 6 ☐	12 − 6 ☐	8 − 2 ☐			
7	7 − 5 ☐	11 − 1 ☐	8 − 6 ☐	9 − 5 ☐	10 − 7 ☐	12 − 5 ☐	11 − 11 ☐		
8	10 − 4 ☐	12 − 3 ☐	9 − 8 ☐	11 − 10 ☐	10 − 1 ☐	12 − 10 ☐	11 − 5 ☐	7 − 7 ☐	
9	12 − 8 ☐	9 − 2 ☐	11 − 2 ☐	9 − 0 ☐	12 − 4 ☐	11 − 4 ☐	10 − 8 ☐	9 − 6 ☐	8 − 3 ☐
10	10 − 5 ☐	12 − 9 ☐	9 − 7 ☐	11 − 3 ☐	8 − 5 ☐	10 − 9 ☐	9 − 1 ☐	12 − 12 ☐	10 − 0 ☐	8 − 4 ☐

GA1392

#									
1	$10 \times 5 = \square$								
2	$9 \times 9 = \square$	$10 \times 2 = \square$							
3	$8 \times 6 = \square$	$11 \times 1 = \square$	$6 \times 5 = \square$						
4	$9 \times 7 = \square$	$4 \times 9 = \square$	$10 \times 3 = \square$	$2 \times 5 = \square$					
5	$11 \times 2 = \square$	$6 \times 8 = \square$	$5 \times 9 = \square$	$12 \times 1 = \square$	$10 \times 9 = \square$				
6	$1 \times 1 = \square$	$5 \times 6 = \square$	$2 \times 2 = \square$	$1 \times 10 = \square$	$6 \times 9 = \square$	$11 \times 0 = \square$			
7	$7 \times 8 = \square$	$12 \times 2 = \square$	$3 \times 8 = \square$	$10 \times 10 = \square$	$5 \times 5 = \square$	$10 \times 6 = \square$	$5 \times 2 = \square$		
8	$10 \times 7 = \square$	$6 \times 4 = \square$	$5 \times 10 = \square$	$9 \times 8 = \square$	$12 \times 0 = \square$	$3 \times 9 = \square$	$8 \times 8 = \square$	$6 \times 3 = \square$	
9	$8 \times 9 = \square$	$6 \times 6 = \square$	$1 \times 12 = \square$	$10 \times 1 = \square$	$4 \times 4 = \square$	$4 \times 6 = \square$	$11 \times 3 = \square$	$9 \times 5 = \square$	$8 \times 7 = \square$
10	$10 \times 8 = \square$	$9 \times 4 = \square$	$7 \times 7 = \square$	$9 \times 6 = \square$	$8 \times 3 = \square$	$10 \times 4 = \square$	$3 \times 3 = \square$	$7 \times 9 = \square$	$3 \times 10 = \square$

Name _____

152

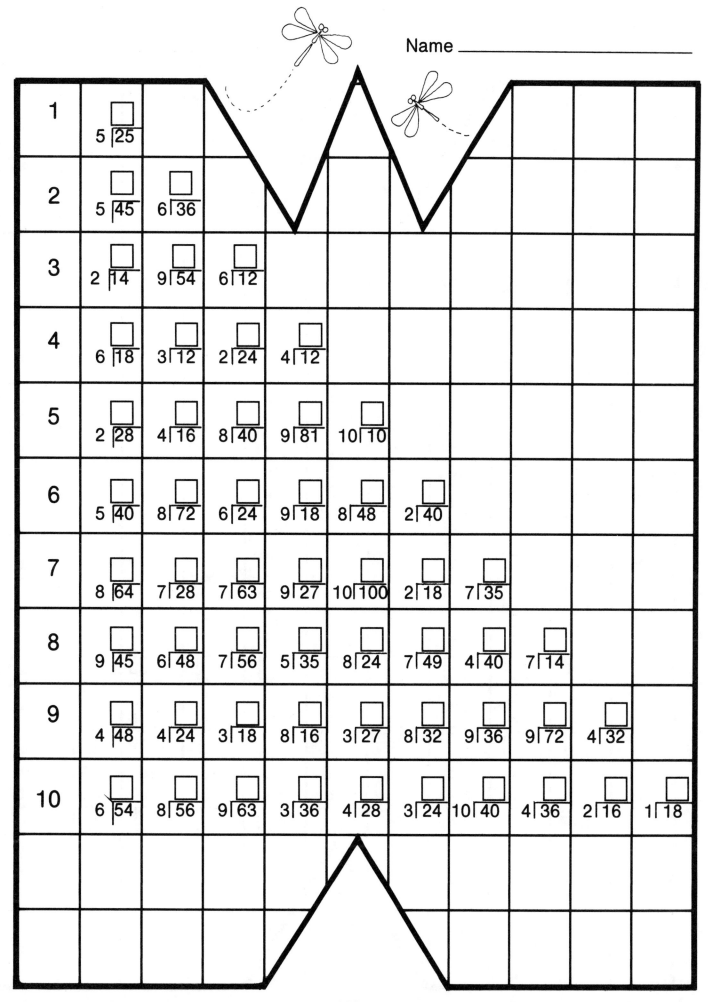

1	☐ 5⟌25								
2	☐ 5⟌45	☐ 6⟌36							
3	☐ 2⟌14	☐ 9⟌54	☐ 6⟌12						
4	☐ 6⟌18	☐ 3⟌12	☐ 2⟌24	☐ 4⟌12					
5	☐ 2⟌28	☐ 4⟌16	☐ 8⟌40	☐ 9⟌81	☐ 10⟌10				
6	☐ 5⟌40	☐ 8⟌72	☐ 6⟌24	☐ 9⟌18	☐ 8⟌48	☐ 2⟌40			
7	☐ 8⟌64	☐ 7⟌28	☐ 7⟌63	☐ 9⟌27	☐ 10⟌100	☐ 2⟌18	☐ 7⟌35		
8	☐ 9⟌45	☐ 6⟌48	☐ 7⟌56	☐ 5⟌35	☐ 8⟌24	☐ 7⟌49	☐ 4⟌40	☐ 7⟌14	
9	☐ 4⟌48	☐ 4⟌24	☐ 3⟌18	☐ 8⟌16	☐ 3⟌27	☐ 8⟌32	☐ 9⟌36	☐ 9⟌72	☐ 4⟌32
10	☐ 6⟌54	☐ 8⟌56	☐ 9⟌63	☐ 3⟌36	☐ 4⟌28	☐ 3⟌24	☐ 10⟌40	☐ 4⟌36	☐ 2⟌16	☐ 1⟌18

GA1392

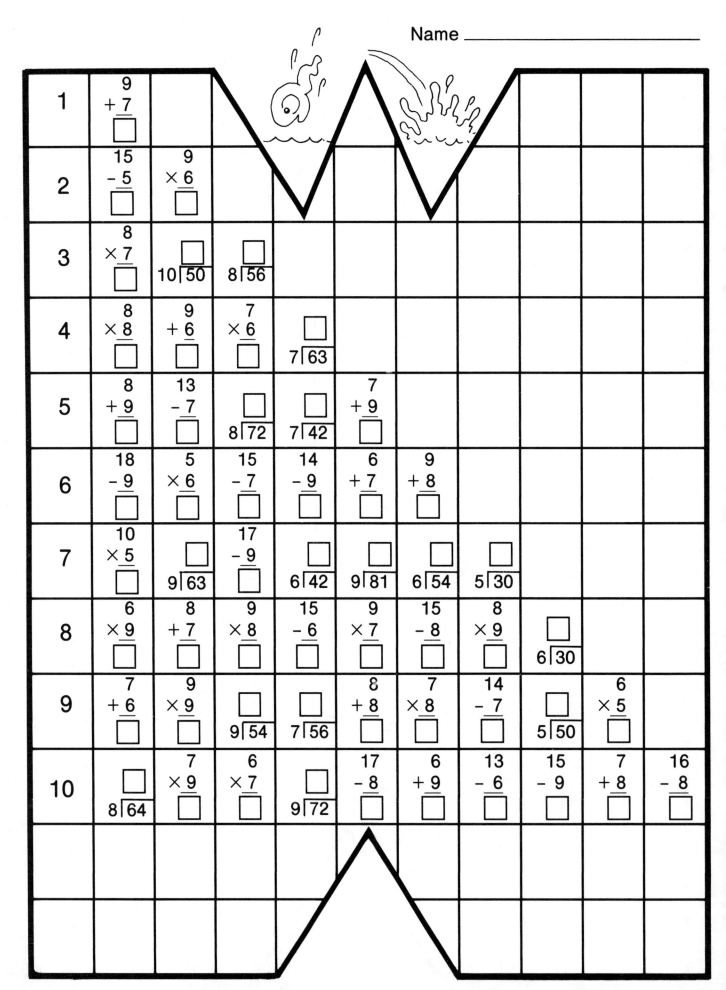

1	9 + 7 □								
2	15 − 5 □	9 × 6 □							
3	8 × 7 □	□ 10)50	□ 8)56						
4	8 × 8 □	9 + 6 □	7 × 6 □	□ 7)63					
5	8 + 9 □	13 − 7 □	□ 8)72	□ 7)42	7 + 9 □				
6	18 − 9 □	5 × 6 □	15 − 7 □	14 − 9 □	6 + 7 □	9 + 8 □			
7	10 × 5 □	□ 9)63	17 − 9 □	□ 6)42	□ 9)81	□ 6)54	□ 5)30		
8	6 × 9 □	8 + 7 □	9 × 8 □	15 − 6 □	9 × 7 □	15 − 8 □	8 × 9 □	□ 6)30	
9	7 + 6 □	9 × 9 □	□ 9)54	□ 7)56	8 + 8 □	7 × 8 □	14 − 7 □	□ 5)50	6 × 5 □
10	□ 8)64	7 × 9 □	6 × 7 □	□ 9)72	17 − 8 □	6 + 9 □	13 − 6 □	15 − 9 □	7 + 8 □	16 − 8 □

Name _____

154

Name _____

This page is for you to create your own daily drill.

155

1	14 + 2 ☐										
2	11 + 0 ☐	12 + 2 ☐									
3	16 + 2 ☐	13 + 5 ☐	15 + 3 ☐								
4	12 + 3 ☐	15 + 2 ☐	10 + 4 ☐	16 + 1 ☐							
5	15 + 5 ☐	11 + 1 ☐	13 + 1 ☐	19 + 0 ☐	18 + 1 ☐						
6	11 + 7 ☐	13 + 6 ☐	14 + 1 ☐	12 + 4 ☐	10 + 5 ☐	11 + 6 ☐					
7	15 + 1 ☐	11 + 4 ☐	16 + 3 ☐	10 + 3 ☐	18 + 2 ☐	11 + 2 ☐	13 + 3 ☐				
8	10 + 0 ☐	13 + 7 ☐	11 + 9 ☐	13 + 0 ☐	14 + 3 ☐	12 + 5 ☐	14 + 4 ☐	13 + 2 ☐			
9	16 + 4 ☐	12 + 0 ☐	18 + 0 ☐	10 + 1 ☐	13 + 4 ☐	17 + 2 ☐	11 + 5 ☐	14 + 6 ☐	20 + 0 ☐		
10	10 + 10 ☐	12 + 6 ☐	11 + 3 ☐	17 + 3 ☐	10 + 2 ☐	12 + 1 ☐	14 + 5 ☐	17 + 1 ☐	12 + 8 ☐	19 + 1 ☐	

156

GA1392

1	15 − 5 □									
2	19 − 3 □	14 − 3 □								
3	12 − 8 □	16 − 2 □	13 − 3 □							
4	14 − 0 □	12 − 2 □	17 − 6 □	16 − 1 □						
5	18 − 4 □	12 − 7 □	15 − 4 □	19 − 4 □	16 − 6 □					
6	14 − 4 □	17 − 7 □	19 − 7 □	15 − 7 □	17 − 5 □	10 − 10 □				
7	18 − 3 □	16 − 5 □	14 − 7 □	13 − 2 □	14 − 1 □	18 − 5 □	12 − 12 □			
8	16 − 0 □	17 − 4 □	12 − 1 □	15 − 3 □	12 − 9 □	16 − 8 □	19 − 6 □	19 − 0 □		
9	13 − 1 □	12 − 6 □	16 − 4 □	14 − 2 □	20 − 1 □	13 − 5 □	15 − 0 □	17 − 2 □	18 − 9 □	
10	17 − 3 □	15 − 2 □	12 − 0 □	20 − 20 □	16 − 3 □	13 − 0 □	13 − 13 □	15 − 1 □	13 − 6 □	18 − 4 □

157

GA1392

1	2 × 10 □										
2	3 × 8 □	11 × 0 □									
3	10 × 4 □	2 × 9 □	9 × 9 □								
4	8 × 4 □	9 × 4 □	12 × 1 □	10 × 3 □							
5	5 × 10 □	10 × 8 □	8 × 7 □	9 × 3 □	3 × 3 □						
6	9 × 8 □	8 × 8 □	5 × 8 □	12 × 2 □	4 × 4 □	10 × 6 □					
7	10 × 1 □	11 × 1 □	9 × 5 □	7 × 9 □	10 × 10 □	9 × 6 □	10 × 0 □				
8	6 × 9 □	7 × 8 □	12 × 0 □	3 × 9 □	4 × 8 □	10 × 9 □	9 × 0 □	8 × 3 □			
9	8 × 5 □	10 × 5 □	7 × 7 □	10 × 2 □	6 × 10 □	0 × 12 □	8 × 6 □	6 × 6 □	6 × 8 □		
10	1 × 12 □	4 × 9 □	8 × 9 □	10 × 7 □	9 × 7 □	5 × 5 □	5 × 9 □	11 × 2 □	9 × 2 □	0 × 9 □	

GA1392

1 □ 7⟌49

2 □ 10⟌50 □ 8⟌40

3 □ 2⟌40 □ 10⟌60 □ 5⟌55

4 □ 11⟌22 □ 4⟌36 □ 7⟌35 □ 8⟌64

5 □ 10⟌70 □ 4⟌44 □ 3⟌24 □ 10⟌30 □ 7⟌56

6 □ 5⟌25 □ 5⟌40 □ 9⟌90 □ 12⟌12 □ 2⟌60 □ 6⟌36

7 □ 2⟌42 □ 8⟌80 □ 2⟌48 □ 9⟌36 □ 8⟌24 □ 5⟌45 □ 8⟌48

8 □ 6⟌30 □ 10⟌100 □ 5⟌35 □ 7⟌77 □ 6⟌60 □ 9⟌81 □ 8⟌56 □ 3⟌36

9 □ 9⟌99 □ 7⟌42 □ 3⟌60 □ 4⟌16 □ 11⟌11 □ 2⟌30 □ 10⟌80 □ 3⟌33 □ 6⟌48

10 □ 10⟌90 □ 8⟌88 □ 2⟌36 □ 3⟌30 □ 7⟌70 □ 4⟌48 □ 6⟌42 □ 10⟌10 □ 2⟌22 □ 5⟌30

GA1392

Name _____

1

$10 \times 8 = \square$

2

$4 \times 10 = \square$ $\quad$ $\square$ $8\overline{)80}$

3

$20 + 4 = \square$ $\quad$ $7 + 8 = \square$ $\quad$ $10 \times 9 = \square$

4

$10 \times 4 = \square$ $\quad$ $15 - 8 = \square$ $\quad$ $8 + 6 = \square$ $\quad$ $16 - 10 = \square$

5

$10 \times 6 = \square$ $\quad$ $20 - 10 = \square$ $\quad$ $\square$ $7\overline{)56}$ $\quad$ $10 + 8 = \square$ $\quad$ $\square$ $9\overline{)90}$

6

$9 + 10 = \square$ $\quad$ $20 - 3 = \square$ $\quad$ $6 \times 8 = \square$ $\quad$ $10 \times 10 = \square$ $\quad$ $\square$ $10\overline{)40}$ $\quad$ $20 \times 1 = \square$

7

$20 \times 3 = \square$ $\quad$ $15 - 7 = \square$ $\quad$ $\square$ $4\overline{)40}$ $\quad$ $18 - 10 = \square$ $\quad$ $\square$ $4\overline{)88}$ $\quad$ $7 \times 8 = \square$ $\quad$ $14 - 10 = \square$

8

$19 - 9 = \square$ $\quad$ $10 + 6 = \square$ $\quad$ $\square$ $10\overline{)90}$ $\quad$ $\square$ $8\overline{)48}$ $\quad$ $\square$ $8\overline{)56}$ $\quad$ $\square$ $10\overline{)100}$ $\quad$ $8 + 10 = \square$ $\quad$ $6 + 8 = \square$

9

$\square$ $10\overline{)60}$ $\quad$ $\square$ $2\overline{)100}$ $\quad$ $18 - 8 = \square$ $\quad$ $8 \times 10 = \square$ $\quad$ $20 \times 2 = \square$ $\quad$ $8 \times 7 = \square$ $\quad$ $10 + 9 = \square$ $\quad$ $8 \times 6 = \square$ $\quad$ $\square$ $2\overline{)40}$

10

$19 - 10 = \square$ $\quad$ $6 \times 10 = \square$ $\quad$ $9 \times 10 = \square$ $\quad$ $8 + 7 = \square$ $\quad$ $\square$ $8\overline{)88}$ $\quad$ $10 + 10 = \square$ $\quad$ $6 + 10 = \square$ $\quad$ $18 - 9 = \square$ $\quad$ $20 \times 4 = \square$ $\quad$ $\square$ $6\overline{)48}$

160

GA1392

This page is for you to create your own daily drill.

161

1	21 + 3 ☐										
2	17 + 3 ☐	19 + 1 ☐									
3	18 + 2 ☐	15 + 4 ☐	23 + 1 ☐								
4	11 + 9 ☐	18 + 4 ☐	16 + 5 ☐	14 + 2 ☐							
5	19 + 5 ☐	15 + 5 ☐	20 + 1 ☐	17 + 7 ☐	20 + 4 ☐						
6	15 + 7 ☐	22 + 2 ☐	17 + 0 ☐	18 + 5 ☐	14 + 1 ☐	13 + 7 ☐					
7	21 + 2 ☐	17 + 1 ☐	19 + 4 ☐	24 + 0 ☐	12 + 8 ☐	19 + 2 ☐	16 + 8 ☐				
8	15 + 6 ☐	20 + 3 ☐	17 + 6 ☐	18 + 6 ☐	22 + 0 ☐	17 + 2 ☐	22 + 1 ☐	14 + 4 ☐			
9	18 + 3 ☐	23 + 0 ☐	16 + 4 ☐	21 + 1 ☐	18 + 0 ☐	16 + 2 ☐	20 + 2 ☐	17 + 4 ☐	18 + 1 ☐		
10	16 + 3 ☐	17 + 5 ☐	19 + 3 ☐	20 + 0 ☐	15 + 9 ☐	19 + 0 ☐	21 + 0 ☐	14 + 3 ☐	15 + 8 ☐	14 + 10 ☐	

162

GA1392

#										
1	17 −7 ☐									
2	18 −8 ☐	15 −3 ☐								
3	24 −0 ☐	19 −7 ☐	14 −4 ☐							
4	16 −10 ☐	15 −4 ☐	24 −4 ☐	16 −6 ☐						
5	17 −6 ☐	22 −0 ☐	16 −3 ☐	19 −9 ☐	23 −3 ☐					
6	15 −2 ☐	19 −6 ☐	17 −5 ☐	19 −10 ☐	18 −6 ☐	14 −3 ☐				
7	23 −2 ☐	18 −7 ☐	24 −3 ☐	16 −5 ☐	18 −10 ☐	13 −13 ☐	14 −0 ☐			
8	18 −5 ☐	21 −1 ☐	15 −5 ☐	18 −3 ☐	24 −2 ☐	19 −5 ☐	22 −1 ☐	19 −2 ☐		
9	19 −8 ☐	17 −4 ☐	17 −10 ☐	15 −10 ☐	13 −10 ☐	12 −12 ☐	23 −1 ☐	22 −2 ☐	13 −0 ☐	
10	16 −4 ☐	13 −2 ☐	24 −1 ☐	19 −4 ☐	13 −3 ☐	23 −0 ☐	13 −1 ☐	17 −3 ☐	19 −3 ☐	12 −2 ☐

163

GA1392

1
$$10 \times 10 = \square$$

2
$$20 \times 1 = \square \qquad 15 \times 2 = \square$$

3
$$6 \times 4 = \square \qquad 10 \times 8 = \square \qquad 11 \times 1 = \square$$

4
$$13 \times 1 = \square \qquad 9 \times 2 = \square \qquad 4 \times 7 = \square \qquad 9 \times 8 = \square$$

5
$$11 \times 0 = \square \qquad 3 \times 8 = \square \qquad 11 \times 3 = \square \qquad 14 \times 1 = \square \qquad 8 \times 8 = \square$$

6
$$2 \times 9 = \square \qquad 10 \times 9 = \square \qquad 6 \times 7 = \square \qquad 12 \times 2 = \square \qquad 5 \times 5 = \square \qquad 7 \times 4 = \square$$

7
$$8 \times 3 = \square \qquad 3 \times 11 = \square \qquad 15 \times 0 = \square \qquad 20 \times 3 = \square \qquad 13 \times 2 = \square \qquad 5 \times 6 = \square \qquad 30 \times 1 = \square$$

8
$$11 \times 2 = \square \qquad 8 \times 9 = \square \qquad 9 \times 10 = \square \qquad 25 \times 1 = \square \qquad 6 \times 6 = \square \qquad 14 \times 2 = \square \qquad 2 \times 12 = \square \qquad 3 \times 5 = \square$$

9
$$7 \times 6 = \square \qquad 9 \times 9 = \square \qquad 11 \times 2 = \square \qquad 14 \times 0 = \square \qquad 8 \times 10 = \square \qquad 12 \times 1 = \square \qquad 7 \times 7 = \square \qquad 4 \times 6 = \square \qquad 1 \times 13 = \square$$

10
$$15 \times 1 = \square \qquad 6 \times 5 = \square \qquad 25 \times 2 = \square \qquad 12 \times 0 = \square \qquad 5 \times 4 = \square \qquad 16 \times 2 = \square \qquad 18 \times 1 = \square \qquad 20 \times 2 = \square \qquad 5 \times 3 = \square \qquad 4 \times 5 = \square$$

GA1392

Name _____

1	$8\overline{)40}$										
2	$7\overline{)28}$	$3\overline{)24}$									
3	$4\overline{)44}$	$10\overline{)60}$	$10\overline{)80}$								
4	$2\overline{)26}$	$7\overline{)49}$	$7\overline{)70}$	$1\overline{)24}$							
5	$3\overline{)36}$	$1\overline{)60}$	$2\overline{)80}$	$10\overline{)50}$	$1\overline{)18}$						
6	$5\overline{)50}$	$4\overline{)28}$	$8\overline{)88}$	$5\overline{)40}$	$9\overline{)27}$	$4\overline{)24}$					
7	$1\overline{)70}$	$5\overline{)55}$	$2\overline{)24}$	$1\overline{)11}$	$8\overline{)80}$	$2\overline{)22}$	$3\overline{)18}$				
8	$7\overline{)77}$	$6\overline{)60}$	$8\overline{)64}$	$12\overline{)24}$	$10\overline{)40}$	$2\overline{)60}$	$2\overline{)100}$	$6\overline{)18}$			
9	$3\overline{)27}$	$4\overline{)80}$	$2\overline{)50}$	$1\overline{)28}$	$10\overline{)70}$	$6\overline{)24}$	$6\overline{)66}$	$4\overline{)32}$	$4\overline{)20}$		
10	$8\overline{)24}$	$9\overline{)18}$	$4\overline{)40}$	$3\overline{)60}$	$3\overline{)33}$	$10\overline{)100}$	$32\overline{)32}$	$24\overline{)24}$	$5\overline{)20}$	$8\overline{)32}$	

165

GA1392

#										
1	19 − 10 = □									
2	9 × 7 = □	8 + 10 = □								
3	7 × 10 = □	□ 2)100	□ 5)40							
4	□ 7)70	7 + 9 = □	18 − 10 = □	11 × 8 = □						
5	□ 8)40	□ 9)90	8 + 11 = □	□ 7)63	□ 11)88					
6	9 + 10 = □	9 + 7 = □	8 × 10 = □	□ 10)70	19 − 9 = □	16 − 7 = □				
7	10 + 8 = □	20 − 2 = □	□ 9)63	□ 10)100	18 − 9 = □	10 + 10 = □	15 − 10 = □			
8	7 × 9 = □	10 × 7 = □	11 + 8 = □	17 − 10 = □	10 + 9 = □	□ 10)50	□ 10)80	15 − 5 = □		
9	□ 4)40	18 − 8 = □	10 × 10 = □	16 − 9 = □	10 + 5 = □	8 × 11 = □	20 − 10 = □	5 × 10 = □	17 − 7 = □	
10	□ 8)88	□ 5)50	10 + 7 = □	10 × 8 = □	□ 10)90	□ 50)50	10 × 5 = □	□ 8)80	5 + 10 = □	7 + 10 = □

GA1392

This page is for you to create your own daily drill.

GA1392

Chain Format

The Chain Format is a unique and exciting approach when considering the idea of creative drill. In this experience, the drill card may be thought of as being a link in a chain. With all of the chain links intact, you will be able to start a chain at any point and completely route through its entire course. This is accomplished by successfully returning to the original starting point.

The drill begins by giving each student a card. The chains all consist of twenty cards. If the class is comprised of more than twenty students, allow two or more class members to share a card. If the class has less than twenty students, some of the participants will have two or more cards. Therefore, the Chain Format is applicable to large or small group instruction.

Once the cards have been distributed, select a student to begin the chain by reading the back of his/her card aloud, and then wait for the next participant to read the only card that could correctly follow the progression. After each participant has the opportunity to read his/her card, there should be a brief pause for the next player to acknowledge his/her place in the chain by reading the next appropriate card. A successful completion of the activity will occur when all of the cards are read in the correct order, and the initial participant is ready to read his/her card for the second time.

Before you begin to successfully complete the chain in the initial attempt, announce to the class that you will be using a stopwatch and that you will be timing their effort. Once the initial attempt at the chain has been completed, announce the time that has been recorded on the stopwatch to the class. "This is your time on the initial try, and it is the time we would like to better."

WAKE UP THE
KIDS, MAUDE!
WE'RE PLAYING
CHAIN
FORMAT!

GA1392

Try this activity with the class two times each day. Keep striving for the fastest recorded time.

When using the chain with five or less students, you might want to spread the twenty chain cards faceup on the table. Then start with any one of the cards and have each participant take turns selecting the next appropriate answer card. A stopwatch will lend much excitement to the activity.

Each chain in this book has specific basic skills listed at the outset.

169

Multiplication Chain

Facts with products to 100

Front

Copy, cut out, color and laminate.

Back

After you cut the number out, record on the back of each card the copy below.

I have 63. Who has 8 × 8?

I have 64. Who has 7 × 5?

I have 35. Who has 10 × 8?

GA1392

Front   Back

80   I have 80. Who has 9 × 9?

81   I have 81. Who has 6 × 7?

42   I have 42. Who has 8 × 6?

48

I have 48. Who has 5 × 8?

40

I have 40. Who has 9 × 8?

72

I have 72. Who has 10 × 10?

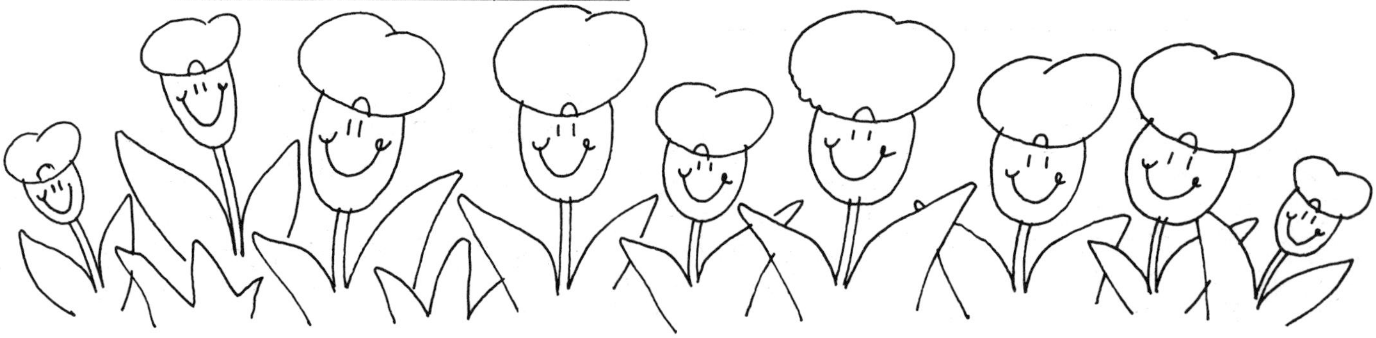

GA1392

Front	Back
100	I have 100. Who has 7 × 7?
49	I have 49. Who has 10 × 5?
50	I have 50. Who has 8 × 7?

GA1392

Front | Back

56

I have 56. Who has 8 × 4?

32

I have 32. Who has 9 × 3?

27

I have 27. Who has 6 × 9?

Front   Back

54   I have 54. Who has 5 × 9?

45   I have 45. Who has 8 × 3?

24   I have 24. Who has 7 × 4?

GA1392

Front   Back

I have 28. Who has 9 × 0?

0

I have 0. Who has 9 × 7?

Re-create the links of each chain on 5 x 7 cards. If you are able to re-create the chain in the actual number form as presented above, you may find that this format will begin with an even greater intensity and motivation.

GA1392

Multiplication Chain

8 and 9 times tables

nine times four.

36!

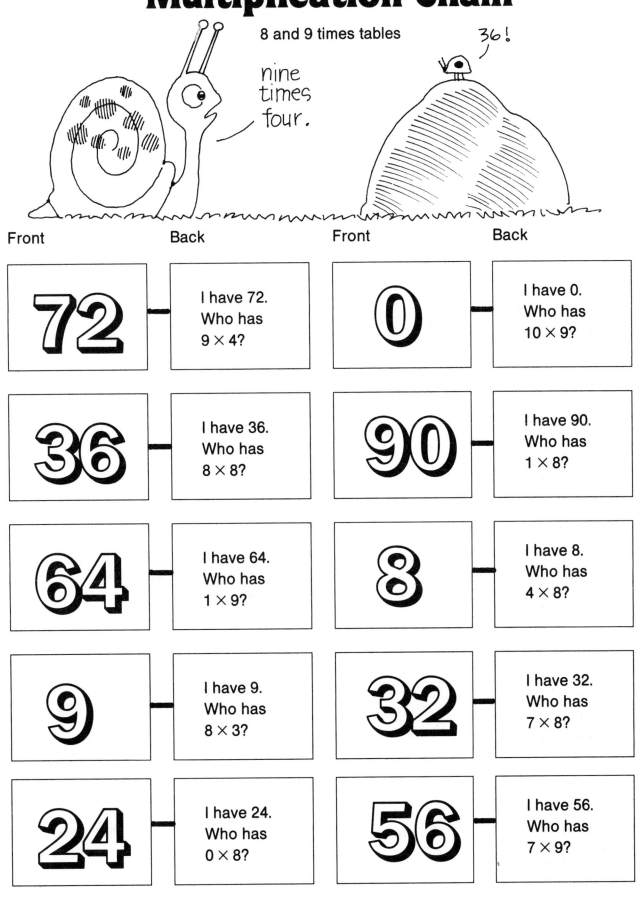

Front	Back	Front	Back
72	I have 72. Who has 9 × 4?	**0**	I have 0. Who has 10 × 9?
36	I have 36. Who has 8 × 8?	**90**	I have 90. Who has 1 × 8?
64	I have 64. Who has 1 × 9?	**8**	I have 8. Who has 4 × 8?
9	I have 9. Who has 8 × 3?	**32**	I have 32. Who has 7 × 8?
24	I have 24. Who has 0 × 8?	**56**	I have 56. Who has 7 × 9?

GA1392

Multiplication Chain

8 and 9 times tables

Front	Back	Front	Back
63	I have 63. Who has 10 × 8?	**40**	I have 40. Who has 5 × 9?
80	I have 80. Who has 2 × 8?	**45**	I have 45. Who has 6 × 8?
16	I have 16. Who has 2 × 9?	**48**	I have 48. Who has 6 × 9?
18	I have 18. Who has 3 × 9?	**54**	I have 54. Who has 9 × 9?
27	I have 27. Who has 5 × 8?	**81**	I have 81. Who has 9 × 8?

GA1392

Multiplication Chain

6 and 7 times tables

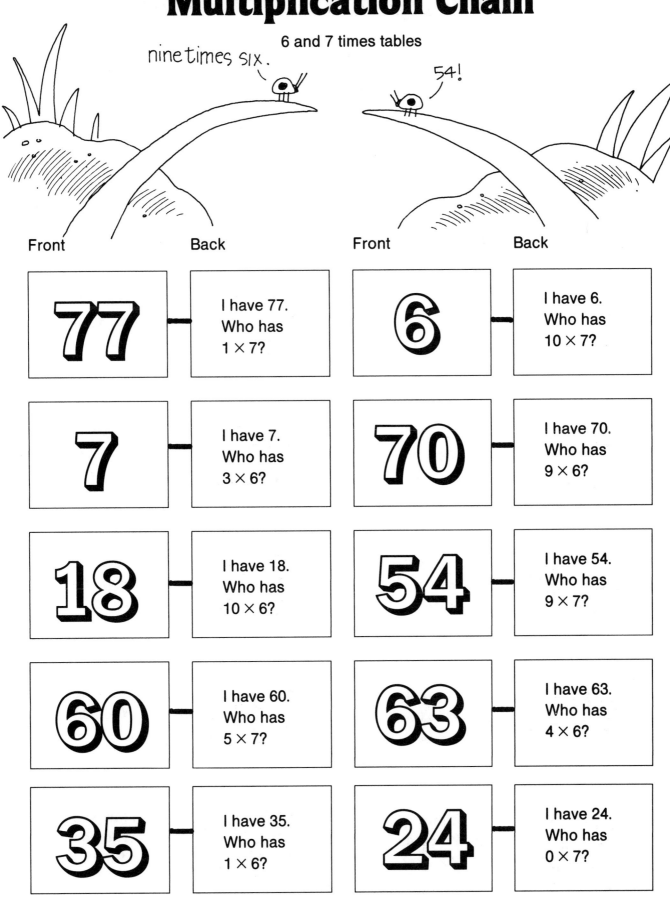

nine times six.

54!

Front   Back   Front   Back

| **77** | I have 77.
Who has
1×7? | **6** | I have 6.
Who has
10×7? |

| **7** | I have 7.
Who has
3×6? | **70** | I have 70.
Who has
9×6? |

| **18** | I have 18.
Who has
10×6? | **54** | I have 54.
Who has
9×7? |

| **60** | I have 60.
Who has
5×7? | **63** | I have 63.
Who has
4×6? |

| **35** | I have 35.
Who has
1×6? | **24** | I have 24.
Who has
0×7? |

GA1392

Multiplication Chain

6 and 7 times tables

Front	Back	Front	Back
0	I have 0. Who has 4×7?	**14**	I have 14. Who has 2×6?
28	I have 28. Who has 8×7?	**12**	I have 12. Who has 3×7?
56	I have 56. Who has 8×6?	**21**	I have 21. Who has 6×6?
48	I have 48. Who has 5×6?	**36**	I have 36. Who has 6×7?
30	I have 30. Who has 2×7?	**42**	I have 42. Who has 11×7?

180

Multiplication Chain

4 and 5 times tables

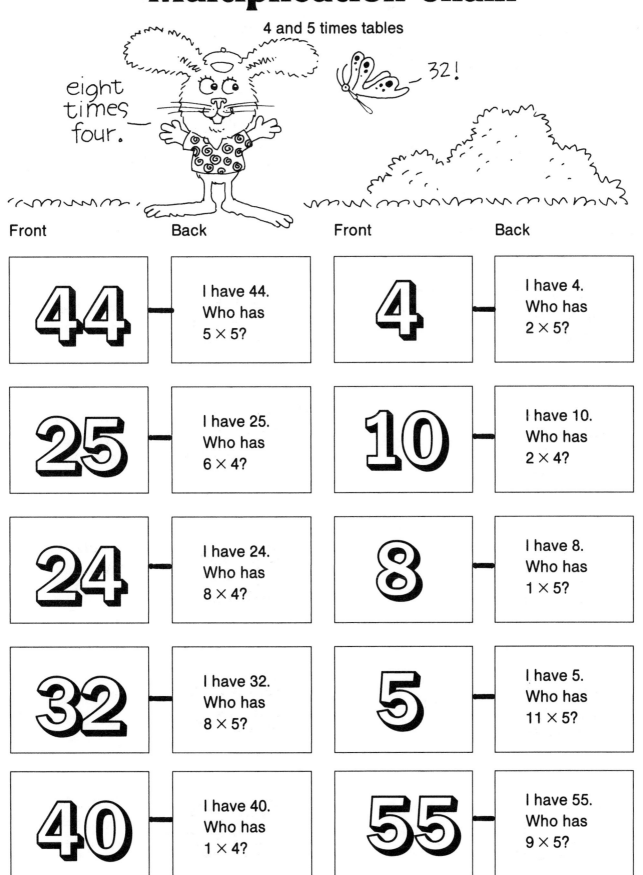

eight times four. — 32!

Front	Back	Front	Back
44	I have 44. Who has 5×5?	**4**	I have 4. Who has 2×5?
25	I have 25. Who has 6×4?	**10**	I have 10. Who has 2×4?
24	I have 24. Who has 8×4?	**8**	I have 8. Who has 1×5?
32	I have 32. Who has 8×5?	**5**	I have 5. Who has 11×5?
40	I have 40. Who has 1×4?	**55**	I have 55. Who has 9×5?

GA1392

Multiplication Chain

4 and 5 times tables

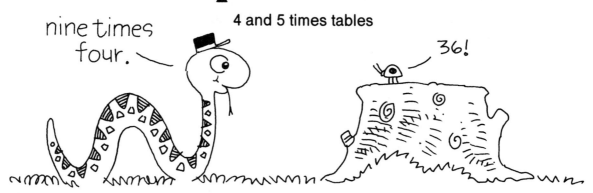

nine times four.

36!

Front	Back	Front	Back
45	I have 45. Who has 9 × 4?	**35**	I have 35. Who has 10 × 5?
36	I have 36. Who has 3 × 4?	**50**	I have 50. Who has 4 × 4?
12	I have 12. Who has 3 × 5?	**16**	I have 16. Who has 4 × 5?
15	I have 15. Who has 7 × 4?	**20**	I have 20. Who has 6 × 5?
28	I have 28. Who has 7 × 5?	**30**	I have 30. Who has 11 × 4?

GA1392

Multiplication Chain

2 and 3 times tables

Front	Back	Front	Back
10	I have 10. Who has 8 × 2?	15	I have 15. Who has 10 × 2?
16	I have 16. Who has 3 × 3?	20	I have 20. Who has 10 × 3?
9	I have 9. Who has 7 × 3?	30	I have 30. Who has 6 × 2?
21	I have 21. Who has 4 × 2?	12	I have 12. Who has 9 × 3?
8	I have 8. Who has 5 × 3?	27	I have 27. Who has 9 × 2?

183

GA1392

Multiplication Chain

2 and 3 times tables

eight times three.

24!

Front	Back	Front	Back
18	I have 18. Who has 2 × 2?	**24**	I have 24. Who has 3 × 2?
4	I have 4. Who has 3 × 1?	**6**	I have 6. Who has 11 × 3?
3	I have 3. Who has 0 × 2?	**33**	I have 33. Who has 11 × 2?
0	I have 0. Who has 7 × 2?	**22**	I have 22. Who has 12 × 2?
14	I have 14. Who has 8 × 3?	**24**	I have 24. Who has 5 × 2?

GA1392

Division Chain

Basic facts

$64 \div 8$

Front	Back	Front	Back

16 — I have 16. Who has $40 \div 10$?

12 — I have 12. Who has $30 \div 10$?

4 — I have 4. Who has $35 \div 7$?

3 — I have 3. Who has $27 \div 3$?

5 — I have 5. Who has $14 \div 2$?

9 — I have 9. Who has $64 \div 8$?

7 — I have 7. Who has $18 \div 3$?

8 — I have 8. Who has $15 \div 1$?

6 — I have 6. Who has $24 \div 2$?

15 — I have 15. Who has $10 \div 10$?

185

GA1392

Division Chain

Basic facts

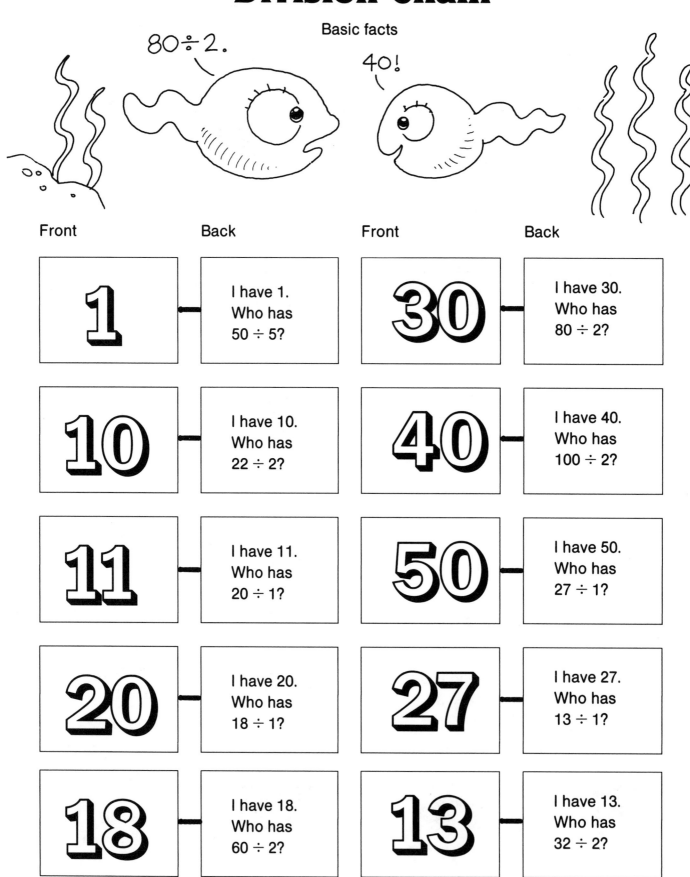

Front   Back   Front   Back

1 — I have 1. Who has 50 ÷ 5?

30 — I have 30. Who has 80 ÷ 2?

10 — I have 10. Who has 22 ÷ 2?

40 — I have 40. Who has 100 ÷ 2?

11 — I have 11. Who has 20 ÷ 1?

50 — I have 50. Who has 27 ÷ 1?

20 — I have 20. Who has 18 ÷ 1?

27 — I have 27. Who has 13 ÷ 1?

18 — I have 18. Who has 60 ÷ 2?

13 — I have 13. Who has 32 ÷ 2?

     GA1392

Addition and Subtraction Chain

Primary facts

Addition and Subtraction Chain

I have 2. Who has 2 more?

Primary facts

4

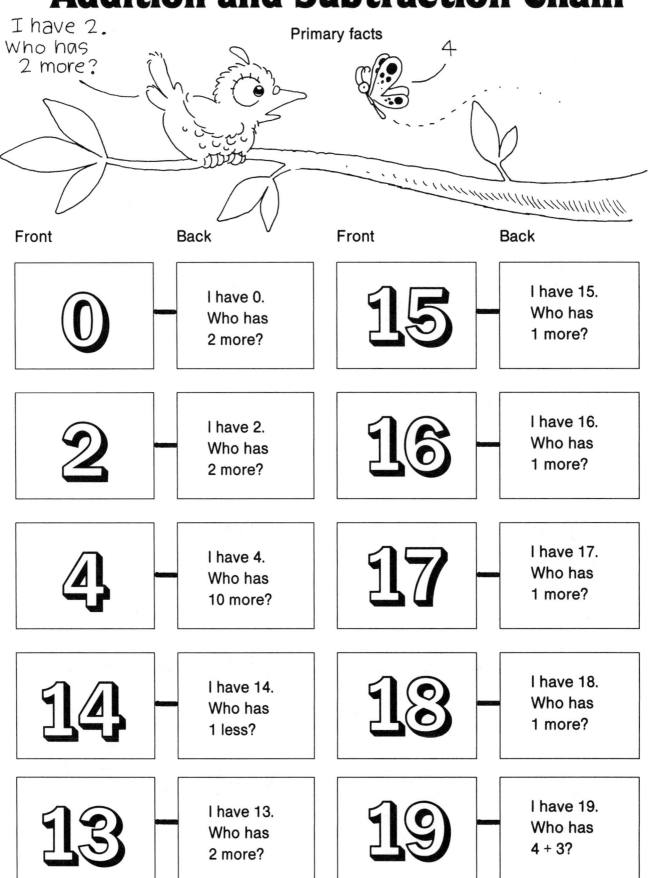

Front	Back	Front	Back
0	I have 0. Who has 2 more?	**15**	I have 15. Who has 1 more?
2	I have 2. Who has 2 more?	**16**	I have 16. Who has 1 more?
4	I have 4. Who has 10 more?	**17**	I have 17. Who has 1 more?
14	I have 14. Who has 1 less?	**18**	I have 18. Who has 1 more?
13	I have 13. Who has 2 more?	**19**	I have 19. Who has 4 + 3?

GA1392

Basic Facts Chain

Intermediate

I have 36.
Who has
14 more?

50!

Front	Back	Front	Back
15	I have 15. Who has its double?	**17**	I have 17. Who has 12 less?
30	I have 30. Who has 1/5 of it?	**5**	I have 5. Who has 31 more?
6	I have 6. Who has 3 times as much?	**36**	I have 36. Who has 14 more?
18	I have 18. Who has 1/2 of it?	**50**	I have 50. Who has 1/5 of it?
9	I have 9. Who has 8 more?	**10**	I have 10. Who has 9 more?

GA1392

Basic Facts Chain

Intermediate

I have 23. Who has 15 less?
8

Front   **Back**   **Front**   **Back**

Front	Back	Front	Back
19	I have 19. Who has its double?	**20**	I have 20. Who has 3 more?
38	I have 38. Who has 2 more?	**23**	I have 23. Who has 15 less?
40	I have 40. Who has $\frac{1}{4}$ of it?	**8**	I have 8. Who has its square plus 1?
10	I have 10. Who has 10 times as much?	**65**	I have 65. Who has 5 less?
100	I have 100. Who has 80 less?	**60**	I have 60. Who has 3×5?

190

GA1392

Odd-Even Number Chain

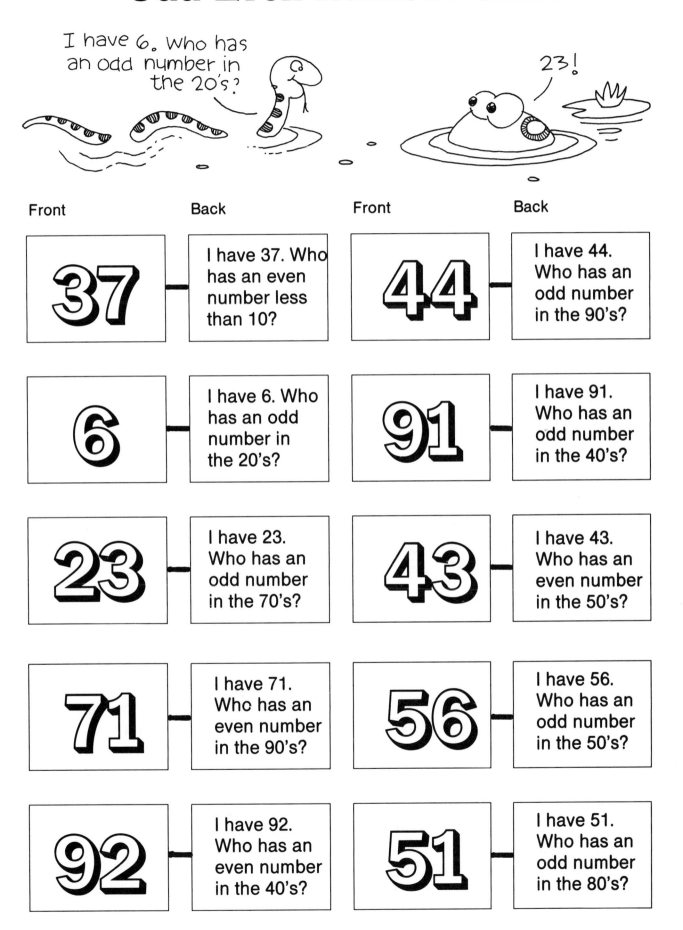

I have 6. Who has an odd number in the 20's?

23!

Front	Back	Front	Back
37	I have 37. Who has an even number less than 10?	44	I have 44. Who has an odd number in the 90's?
6	I have 6. Who has an odd number in the 20's?	91	I have 91. Who has an odd number in the 40's?
23	I have 23. Who has an odd number in the 70's?	43	I have 43. Who has an even number in the 50's?
71	I have 71. Who has an even number in the 90's?	56	I have 56. Who has an odd number in the 50's?
92	I have 92. Who has an even number in the 40's?	51	I have 51. Who has an odd number in the 80's?

GA1392

Odd-Even Number Chain

I have 19. who has an even number in the 20's.

26!

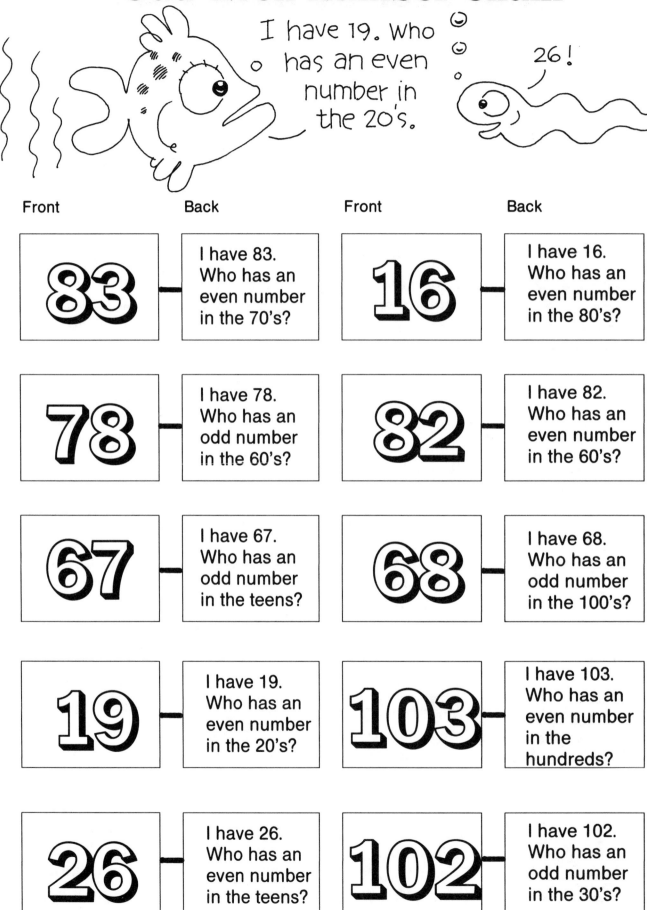

Front	Back	Front	Back
83	I have 83. Who has an even number in the 70's?	**16**	I have 16. Who has an even number in the 80's?
78	I have 78. Who has an odd number in the 60's?	**82**	I have 82. Who has an even number in the 60's?
67	I have 67. Who has an odd number in the teens?	**68**	I have 68. Who has an odd number in the 100's?
19	I have 19. Who has an even number in the 20's?	**103**	I have 103. Who has an even number in the hundreds?
26	I have 26. Who has an even number in the teens?	**102**	I have 102. Who has an odd number in the 30's?

192

R.N.C. (Really Neat Chain)

I have 72. Who has 8 more?

80!

Front	Back	Front	Back
36	I have 36. Who has its double?	**10**	I have 10. Who has ½ of it?
72	I have 72. Who has 8 more?	**5**	I have 5. Who has ½ of it?
80	I have 80. Who has ½ of it?	**2½**	I have 2½. Who has ½ more?
40	I have 40. Who has ½ of it?	**3**	I have 3. Who has 10 times as much?
20	I have 20. Who has ½ of it?	**30**	I have 30. Who has 10 times as much?

193

GA1392

R.N.C. (Really Neat Chain)

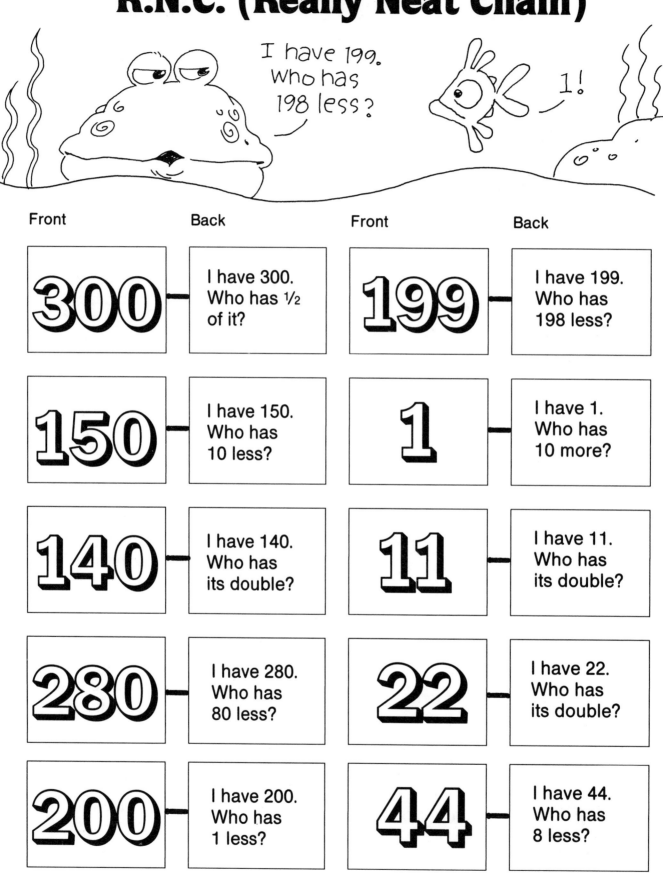

Front	Back	Front	Back
300	I have 300. Who has ½ of it?	199	I have 199. Who has 198 less?
150	I have 150. Who has 10 less?	1	I have 1. Who has 10 more?
140	I have 140. Who has its double?	11	I have 11. Who has its double?
280	I have 280. Who has 80 less?	22	I have 22. Who has its double?
200	I have 200. Who has 1 less?	44	I have 44. Who has 8 less?

Chain for Doubles, Plus 1

Addition

Front	Back	Front	Back
12	I have 12. Who has 6 + 7?	**19**	I have 19. Who has 4 + 4?
13	I have 13. Who has 10 + 10?	**8**	I have 8. Who has 4 + 5?
20	I have 20. Who has 10 + 11?	**9**	I have 9. Who has 7 + 7?
21	I have 21. Who has 9 + 9?	**14**	I have 14. Who has 7 + 8?
18	I have 18. Who has 9 + 10?	**15**	I have 15. Who has 5 + 5?

195

GA1392

Chain for Doubles, Plus 1

Addition

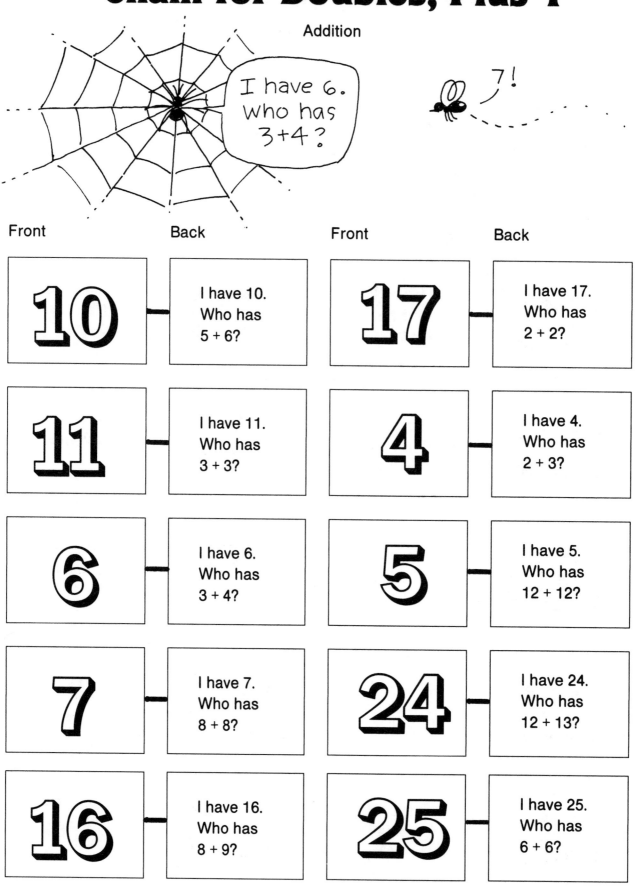

Front   Back   Front   Back

10 — I have 10. Who has 5 + 6?

17 — I have 17. Who has 2 + 2?

11 — I have 11. Who has 3 + 3?

4 — I have 4. Who has 2 + 3?

6 — I have 6. Who has 3 + 4?

5 — I have 5. Who has 12 + 12?

7 — I have 7. Who has 8 + 8?

24 — I have 24. Who has 12 + 13?

16 — I have 16. Who has 8 + 9?

25 — I have 25. Who has 6 + 6?

196

GA1392

Measurement Chain

I have 2 weeks.
Who has another
name for 1 foot?

12
inches.

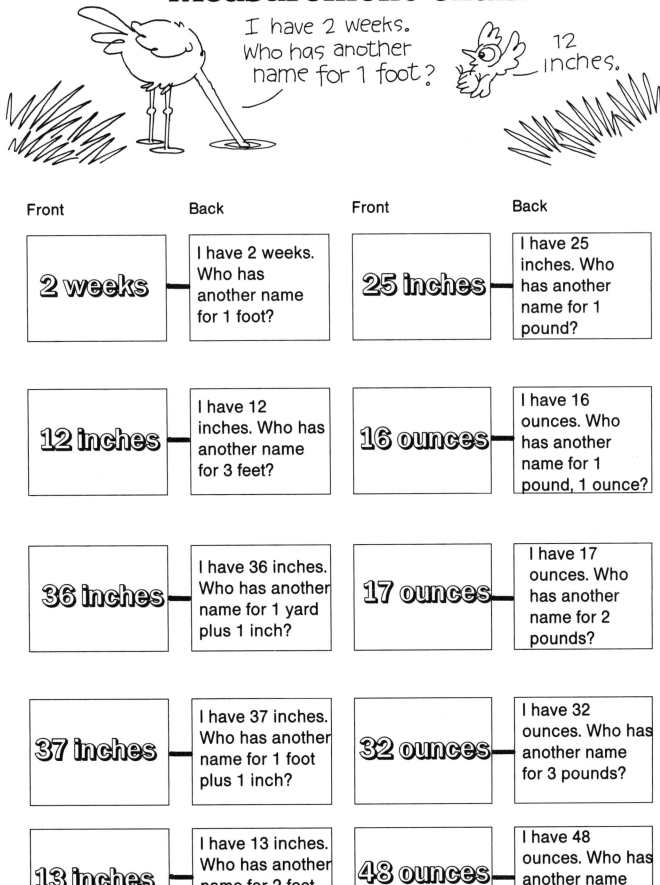

Front	Back	Front	Back
2 weeks	I have 2 weeks. Who has another name for 1 foot?	**25 inches**	I have 25 inches. Who has another name for 1 pound?
12 inches	I have 12 inches. Who has another name for 3 feet?	**16 ounces**	I have 16 ounces. Who has another name for 1 pound, 1 ounce?
36 inches	I have 36 inches. Who has another name for 1 yard plus 1 inch?	**17 ounces**	I have 17 ounces. Who has another name for 2 pounds?
37 inches	I have 37 inches. Who has another name for 1 foot plus 1 inch?	**32 ounces**	I have 32 ounces. Who has another name for 3 pounds?
13 inches	I have 13 inches. Who has another name for 2 feet plus 1 inch?	**48 ounces**	I have 48 ounces. Who has another name for 3 pounds, 3 ounces?

GA1392

Measurement Chain

I have 74 inches. Who has another name for 1 year?

12 months!

Front | Back | Front | Back

51 ounces — I have 51 ounces. Who has another name for 2 yards?

1 month — I have 1 month. Who has another name for 2 years?

72 inches — I have 72 inches. Who has another name for 2 yards, 2 inches?

24 months — I have 24 months. Who has another name for 2 years, 2 months?

74 inches — I have 74 inches. Who has another name for 1 year?

26 months — I have 26 months. Who has another name for 3 years?

12 months — I have 12 months. Who has another name for ½ year?

36 months — I have 36 months. Who has another name for 1 week?

6 months — I have 6 months. Who has another name for 31 days?

7 days — I have 7 days. Who has another name for 14 days?

GA1392

Fraction Chain

I have 6. Who has 3 times as much?

18!

Front	Back	Front	Back
$\frac{1}{4}$	I have $\frac{1}{4}$. Who has $\frac{1}{4} + \frac{1}{4}$?	**3**	I have 3. Who has $3 + \frac{1}{4}$?
$\frac{1}{2}$	I have $\frac{1}{2}$. Who has $\frac{1}{2} + \frac{1}{2}$?	**3$\frac{1}{4}$**	I have $3\frac{1}{4}$. Who has $3\frac{1}{4} + \frac{1}{4}$?
1	I have 1 whole. Who has $1 + \frac{1}{4}$?	**3$\frac{1}{2}$**	I have $3\frac{1}{2}$. Who has its double?
1$\frac{1}{4}$	I have $1\frac{1}{4}$. Who has $1\frac{1}{4} + \frac{1}{4}$?	**7**	I have 7. Who has $7 - 7$?
1$\frac{1}{2}$	I have $1\frac{1}{2}$. Who has $1\frac{1}{2} + 1\frac{1}{2}$?	**0**	I have 0. Who has $\frac{1}{8} + \frac{4}{8}$?

GA1392

Fraction Chain

I have 5.
Who has
5 + 5 ?

10!

Front	Back	Front	Back
$\dfrac{5}{8}$	I have $\dfrac{5}{8}$. Who has $\dfrac{5}{8} + \dfrac{2}{8}$?	**5**	I have 5. Who has 5 + 5 ?
$\dfrac{7}{8}$	I have $\dfrac{7}{8}$. Who has $1 + \dfrac{7}{8}$?	**10**	I have 10. Who has $10 + \dfrac{1}{3}$?
$1\dfrac{7}{8}$	I have $1\dfrac{7}{8}$. Who has $1\dfrac{7}{8} + \dfrac{1}{8}$?	$10\dfrac{1}{3}$	I have $10\dfrac{1}{3}$. Who has $10\dfrac{1}{3} + \dfrac{1}{3}$?
2	I have 2. Who has $2 + \dfrac{1}{2}$ more?	$10\dfrac{2}{3}$	I have $10\dfrac{2}{3}$. Who has $10\dfrac{2}{3} + \dfrac{1}{3}$?
$2\dfrac{1}{2}$	I have $2\dfrac{1}{2}$. Who has $2\dfrac{1}{2} + 2\dfrac{1}{2}$?	**11**	I have 11. Who has 1 whole minus $\dfrac{3}{4}$?

200

GA1392

Basic Division and Fraction Chain

Multiplication

I have 15. Who has 1/5 of it?

3!

Front	Back	Front	Back

50 — I have 50. Who has 1/10 of it?

6 — I have 6. Who has 3 times as much?

5 — I have 5. Who has 3 times as much?

18 — I have 18. Who has 1/2 of it?

15 — I have 15. Who has 1/5 of it?

9 — I have 9. Who has 5 times as much?

3 — I have 3. Who has 10 times as much?

45 — I have 45. Who has its double?

30 — I have 30. Who has 1/5 of it?

90 — I have 90. Who has 1/9 of it?

GA1392

Basic Division and Fraction Chain

Multiplication

Front	Back	Front	Back
10	I have 10. Who has it's double?	**2**	I have 2. Who has ½ of it?
20	I have 20. Who has its double?	**1**	I have 1. Who has ½ of it?
40	I have 40. Who has ⅕ of it?	**½**	I have $\frac{1}{2}$. Who has $6\frac{1}{2}$ more?
8	I have 8. Who has ½ of it?	**7**	I have 7. Who has its double?
4	I have 4. Who has ½ of it?	**14**	I have 14, Who has 36 more?

GA1392

The Money Chain

Pennies, nickels, dimes, quarters

I have $.20. Who has 3 nickels?

$.15!

Front	Back	Front	Back
 $.51	I have $.51. Who has 2 dimes?	$.30	I have $.30. Who has 3 nickels and 3 dimes?
 $.20	I have $.20. Who has 3 nickels?	$.45	I have $.45. Who has 1 quarter and 1 penny?
 $.15	I have $.15. Who has 3 pennies?	$.26	I have $.26. Who has 3 quarters?
 $.03	I have $.03. Who has 4 dimes?	$.75	I have $.75. Who has 2 quarters and 2 nickels?
 $.40	I have $.40. Who has 6 nickels?	 $.60	I have $.60. Who has 1 quarter and 1 dime?

GA1392

The Money Chain

Pennies, nickels, dimes, quarters

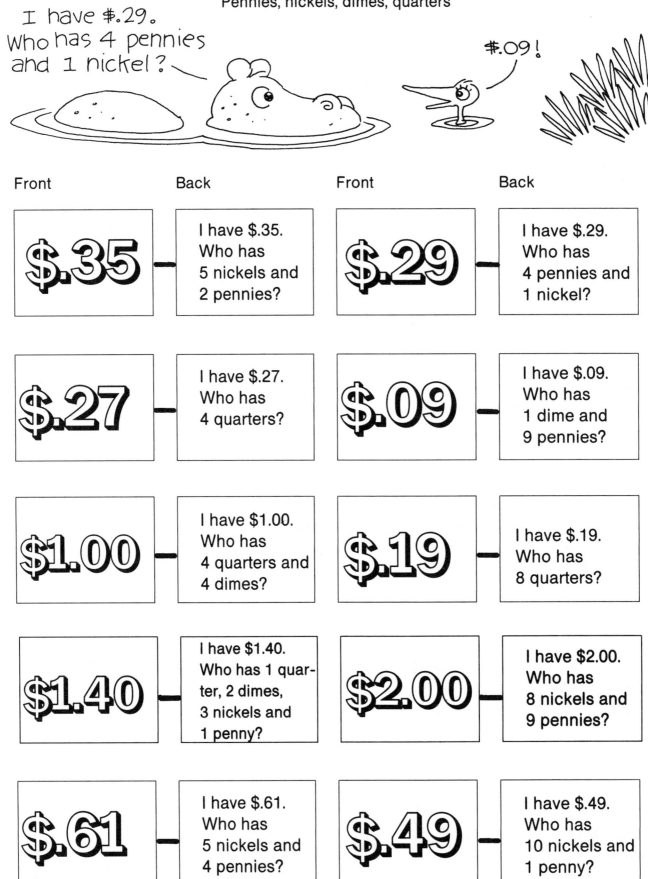

I have $.29.
Who has 4 pennies
and 1 nickel?

$.09!

Front	Back	Front	Back
$.35	I have $.35. Who has 5 nickels and 2 pennies?	**$.29**	I have $.29. Who has 4 pennies and 1 nickel?
$.27	I have $.27. Who has 4 quarters?	**$.09**	I have $.09. Who has 1 dime and 9 pennies?
$1.00	I have $1.00. Who has 4 quarters and 4 dimes?	**$.19**	I have $.19. Who has 8 quarters?
$1.40	I have $1.40. Who has 1 quarter, 2 dimes, 3 nickels and 1 penny?	**$2.00**	I have $2.00. Who has 8 nickels and 9 pennies?
$.61	I have $.61. Who has 5 nickels and 4 pennies?	**$.49**	I have $.49. Who has 10 nickels and 1 penny?

GA1392

The Time Chain

Hour, half-hour, quarter-hour

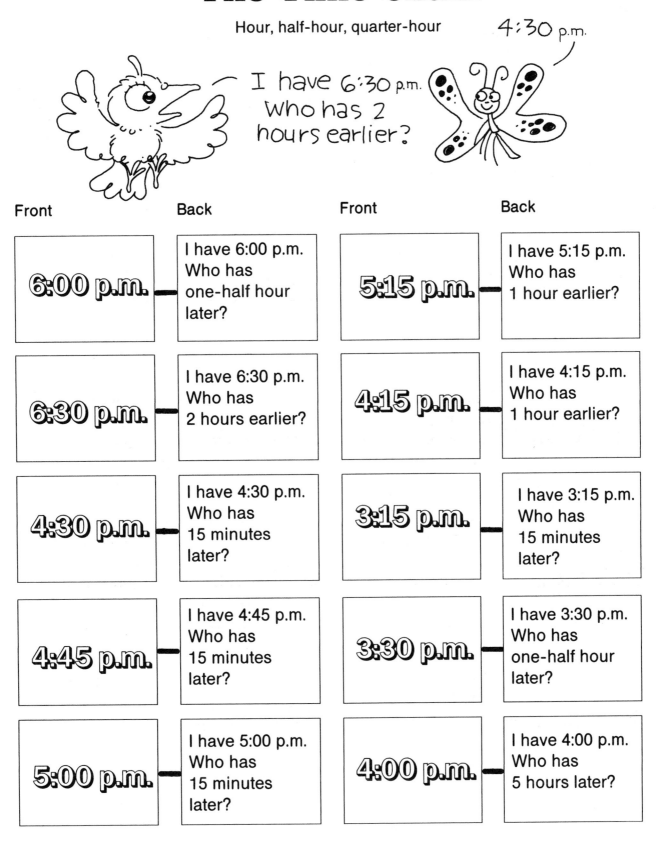

I have 6:30 p.m. Who has 2 hours earlier?

4:30 p.m.

Front	Back	Front	Back
6:00 p.m.	I have 6:00 p.m. Who has one-half hour later?	5:15 p.m.	I have 5:15 p.m. Who has 1 hour earlier?
6:30 p.m.	I have 6:30 p.m. Who has 2 hours earlier?	4:15 p.m.	I have 4:15 p.m. Who has 1 hour earlier?
4:30 p.m.	I have 4:30 p.m. Who has 15 minutes later?	3:15 p.m.	I have 3:15 p.m. Who has 15 minutes later?
4:45 p.m.	I have 4:45 p.m. Who has 15 minutes later?	3:30 p.m.	I have 3:30 p.m. Who has one-half hour later?
5:00 p.m.	I have 5:00 p.m. Who has 15 minutes later?	4:00 p.m.	I have 4:00 p.m. Who has 5 hours later?

GA1392

The Time Chain

Hour, half-hour, quarter-hour

I have 8:15 p.m.
Who has 15
minutes earlier?

8:00 p.m.

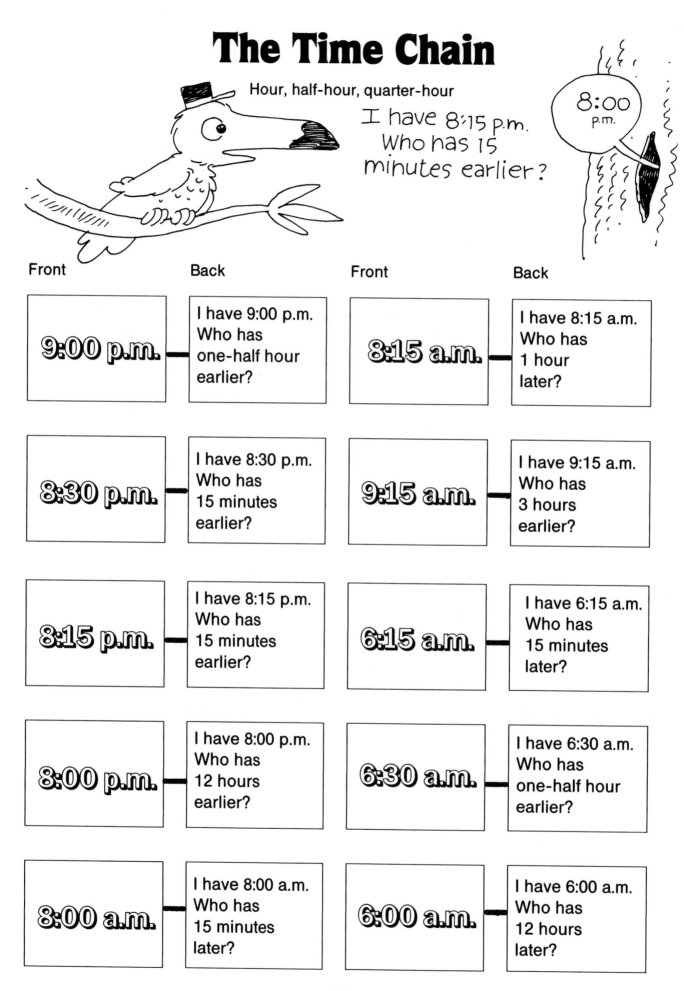

Front	Back	Front	Back
9:00 p.m.	I have 9:00 p.m. Who has one-half hour earlier?	8:15 a.m.	I have 8:15 a.m. Who has 1 hour later?
8:30 p.m.	I have 8:30 p.m. Who has 15 minutes earlier?	9:15 a.m.	I have 9:15 a.m. Who has 3 hours earlier?
8:15 p.m.	I have 8:15 p.m. Who has 15 minutes earlier?	6:15 a.m.	I have 6:15 a.m. Who has 15 minutes later?
8:00 p.m.	I have 8:00 p.m. Who has 12 hours earlier?	6:30 a.m.	I have 6:30 a.m. Who has one-half hour earlier?
8:00 a.m.	I have 8:00 a.m. Who has 15 minutes later?	6:00 a.m.	I have 6:00 a.m. Who has 12 hours later?

206

GA1392

The Time Chain

To the hour, half-hour

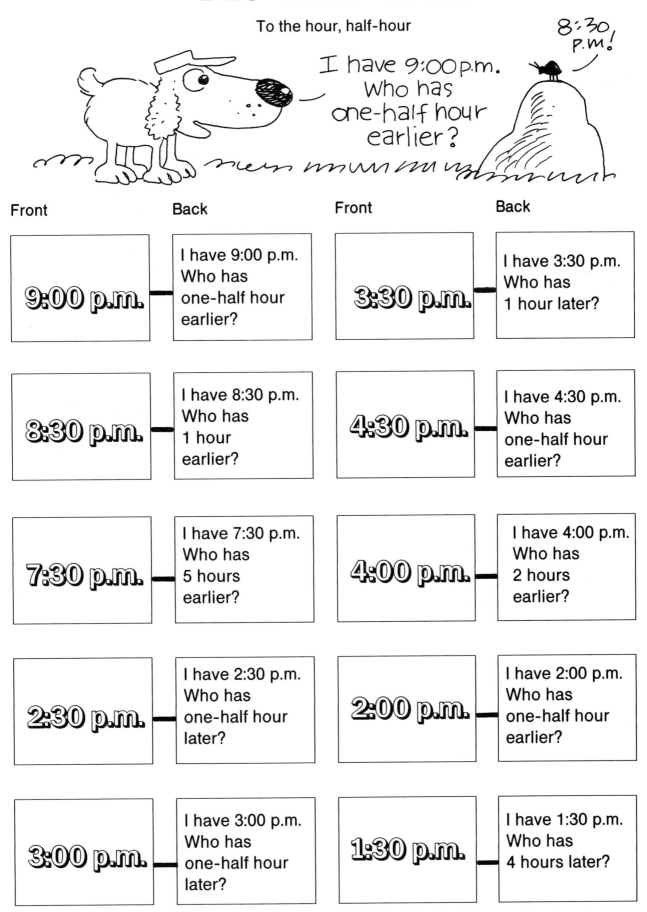

I have 9:00 p.m. Who has one-half hour earlier?

8:30 p.m.!

Front	Back	Front	Back
9:00 p.m.	I have 9:00 p.m. Who has one-half hour earlier?	3:30 p.m.	I have 3:30 p.m. Who has 1 hour later?
8:30 p.m.	I have 8:30 p.m. Who has 1 hour earlier?	4:30 p.m.	I have 4:30 p.m. Who has one-half hour earlier?
7:30 p.m.	I have 7:30 p.m. Who has 5 hours earlier?	4:00 p.m.	I have 4:00 p.m. Who has 2 hours earlier?
2:30 p.m.	I have 2:30 p.m. Who has one-half hour later?	2:00 p.m.	I have 2:00 p.m. Who has one-half hour earlier?
3:00 p.m.	I have 3:00 p.m. Who has one-half hour later?	1:30 p.m.	I have 1:30 p.m. Who has 4 hours later?

GA1392

The Time Chain

To the hour, half-hour

I have 10:00 p.m. who has 1 hour later?

11:00 p.m.

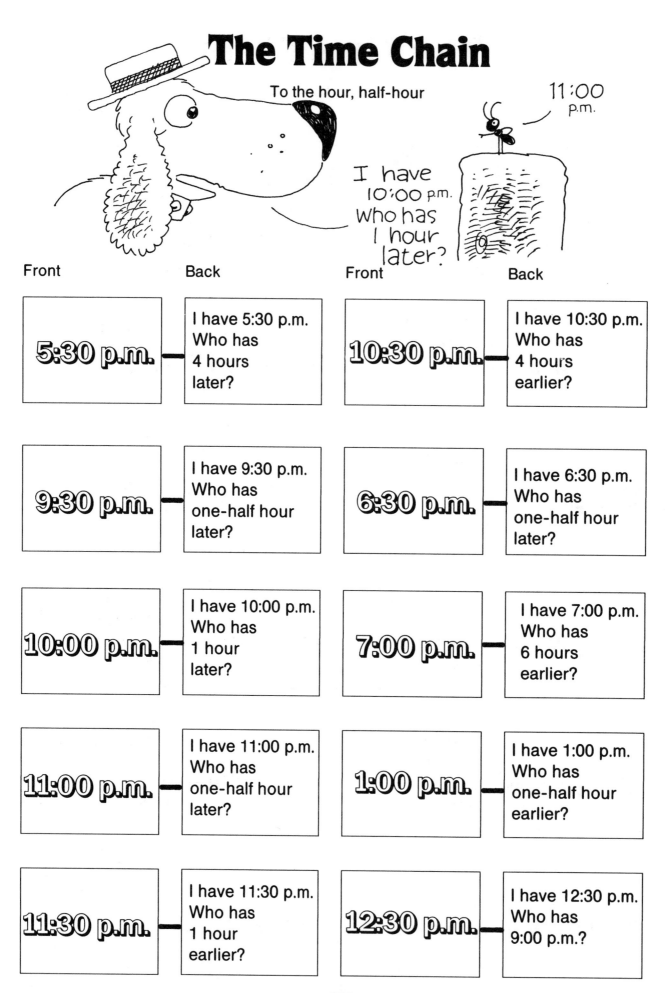

Front	Back	Front	Back
5:30 p.m.	I have 5:30 p.m. Who has 4 hours later?	10:30 p.m.	I have 10:30 p.m. Who has 4 hours earlier?
9:30 p.m.	I have 9:30 p.m. Who has one-half hour later?	6:30 p.m.	I have 6:30 p.m. Who has one-half hour later?
10:00 p.m.	I have 10:00 p.m. Who has 1 hour later?	7:00 p.m.	I have 7:00 p.m. Who has 6 hours earlier?
11:00 p.m.	I have 11:00 p.m. Who has one-half hour later?	1:00 p.m.	I have 1:00 p.m. Who has one-half hour earlier?
11:30 p.m.	I have 11:30 p.m. Who has 1 hour earlier?	12:30 p.m.	I have 12:30 p.m. Who has 9:00 p.m.?

208

GA1392

The Time Chain

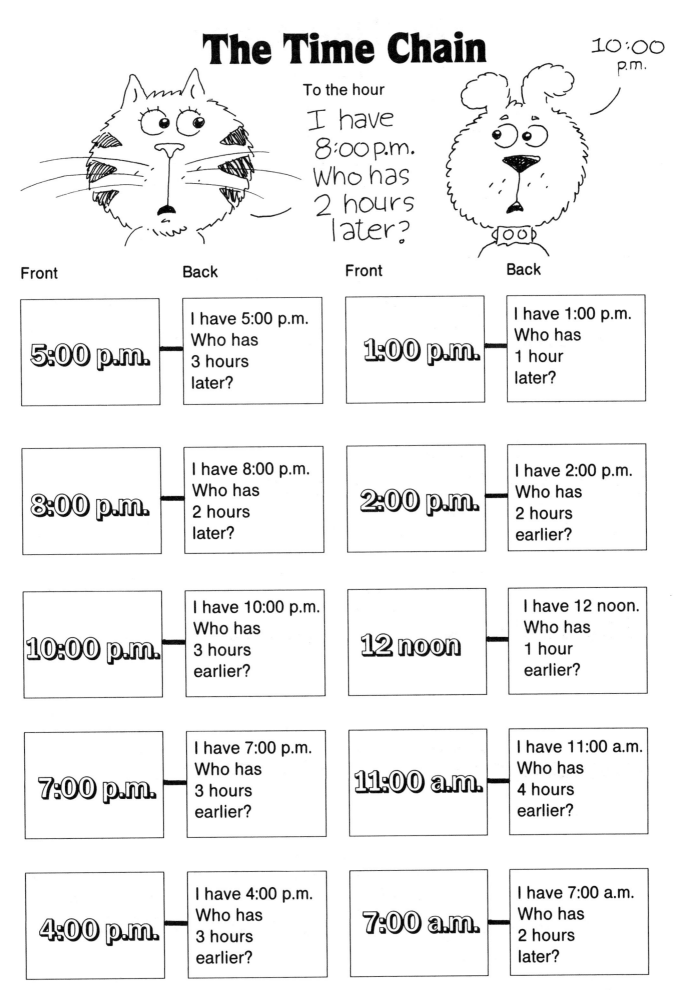

10:00 p.m.

To the hour

I have 8:00 p.m. Who has 2 hours later?

Front	Back	Front	Back
5:00 p.m.	I have 5:00 p.m. Who has 3 hours later?	1:00 p.m.	I have 1:00 p.m. Who has 1 hour later?
8:00 p.m.	I have 8:00 p.m. Who has 2 hours later?	2:00 p.m.	I have 2:00 p.m. Who has 2 hours earlier?
10:00 p.m.	I have 10:00 p.m. Who has 3 hours earlier?	12 noon	I have 12 noon. Who has 1 hour earlier?
7:00 p.m.	I have 7:00 p.m. Who has 3 hours earlier?	11:00 a.m.	I have 11:00 a.m. Who has 4 hours earlier?
4:00 p.m.	I have 4:00 p.m. Who has 3 hours earlier?	7:00 a.m.	I have 7:00 a.m. Who has 2 hours later?

GA1392

The Time Chain

To the hour

I have 10:00 a.m. Who has 5 hours earlier?

5:00 a.m.!

Front	Back	Front	Back
9:00 a.m.	I have 9:00 a.m. Who has 1 hour later?	9:00 p.m.	I have 9:00 p.m. Who has 2 hours later?
10:00 a.m.	I have 10:00 a.m. Who has 5 hours earlier?	11:00 p.m.	I have 11:00 p.m. Who has 2 hours later?
5:00 a.m.	I have 5:00 a.m. Who has 1 hour later?	1:00 a.m.	I have 1:00 a.m. Who has 1 hour later?
6:00 a.m.	I have 6:00 a.m. Who has 12 hours later?	2:00 a.m.	I have 2:00 a.m. Who has 2 hours later?
6:00 p.m.	I have 6:00 p.m. Who has 3 hours later?	4:00 a.m.	I have 4:00 a.m. Who has 5:00 pm?

210

GA1392

Chain for Addition and Subtraction Without Regrouping

Facts from 7-49

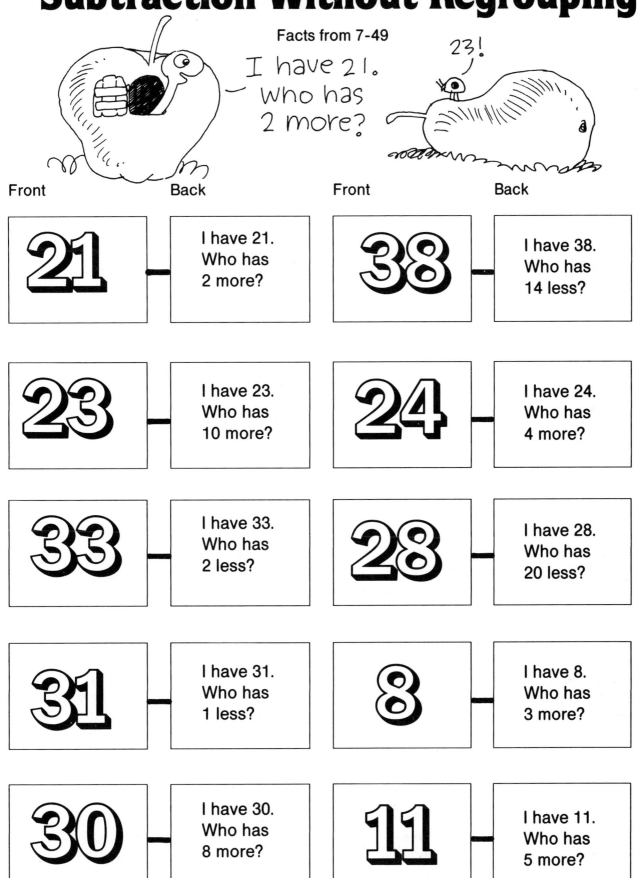

Front   Back   Front   Back

21 — I have 21. Who has 2 more?

38 — I have 38. Who has 14 less?

23 — I have 23. Who has 10 more?

24 — I have 24. Who has 4 more?

33 — I have 33. Who has 2 less?

28 — I have 28. Who has 20 less?

31 — I have 31. Who has 1 less?

8 — I have 8. Who has 3 more?

30 — I have 30. Who has 8 more?

11 — I have 11. Who has 5 more?

211

Chain for Addition and Subtraction Without Regrouping

Facts from 7-49

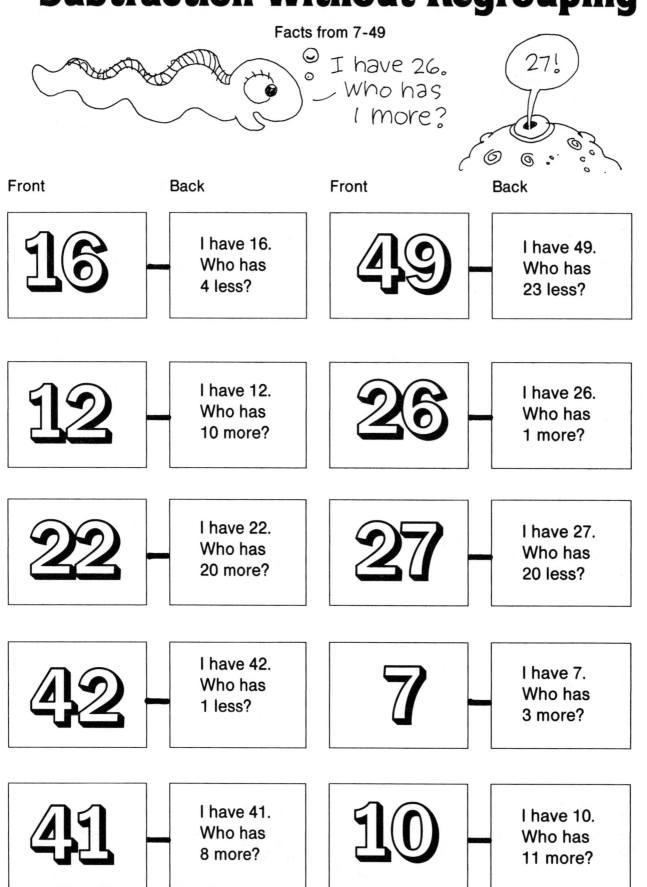

Front	Back	Front	Back
16	I have 16. Who has 4 less?	49	I have 49. Who has 23 less?
12	I have 12. Who has 10 more?	26	I have 26. Who has 1 more?
22	I have 22. Who has 20 more?	27	I have 27. Who has 20 less?
42	I have 42. Who has 1 less?	7	I have 7. Who has 3 more?
41	I have 41. Who has 8 more?	10	I have 10. Who has 11 more?

GA1392

Temperature Chain

I have 80°.
who has
30°
cooler?

50°

Front — Back

Front	Back
70⁰	I have 70⁰. Who has 10⁰ warmer?
80⁰	I have 80⁰. Who has 30⁰ cooler?
50⁰	I have 50⁰. Who has 2⁰ warmer?
52⁰	I have 52⁰. Who has 20⁰ warmer?
72⁰	I have 72⁰. Who has 42⁰ cooler?

Front	Back
30⁰	I have 30⁰. Who has 9⁰ cooler?
21⁰	I have 21⁰. Who has 3⁰ warmer?
24⁰	I have 24⁰. Who has 23⁰ cooler?
1⁰	I have 1⁰. Who has 1⁰ cooler?
0⁰	I have 0⁰. Who has 1⁰ cooler?

GA1392

Temperature Chain

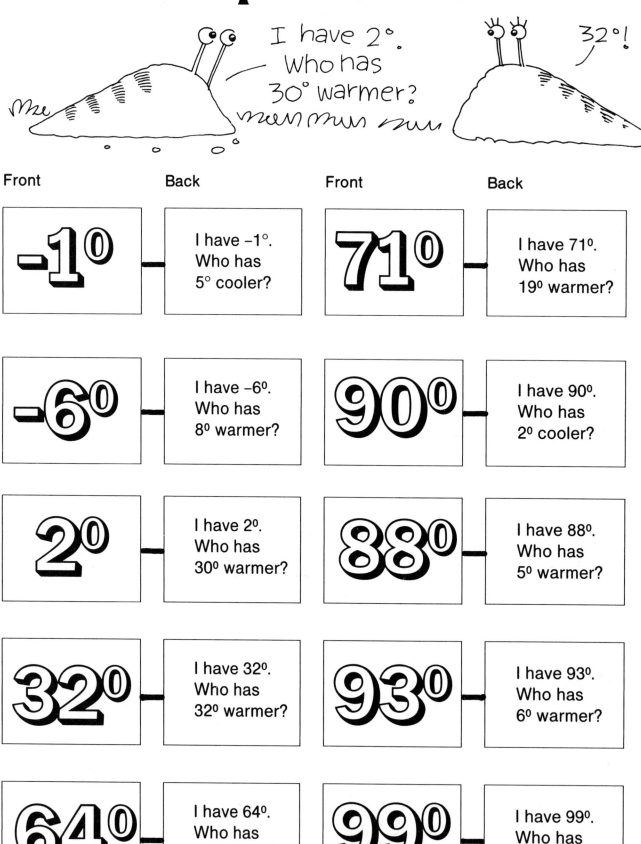

I have 2°. Who has 30° warmer?

32°!

Me

Front	Back	Front	Back
-1⁰	I have –1°. Who has 5° cooler?	**71⁰**	I have 71⁰. Who has 19⁰ warmer?
-6⁰	I have –6⁰. Who has 8⁰ warmer?	**90⁰**	I have 90⁰. Who has 2⁰ cooler?
2⁰	I have 2⁰. Who has 30⁰ warmer?	**88⁰**	I have 88⁰. Who has 5⁰ warmer?
32⁰	I have 32⁰. Who has 32⁰ warmer?	**93⁰**	I have 93⁰. Who has 6⁰ warmer?
64⁰	I have 64⁰. Who has 7⁰ warmer?	**99⁰**	I have 99⁰. Who has 29⁰ cooler?

GA1392

Fraction Chain: Addition, Subtraction of Units, Eighths, Quarters and Halves

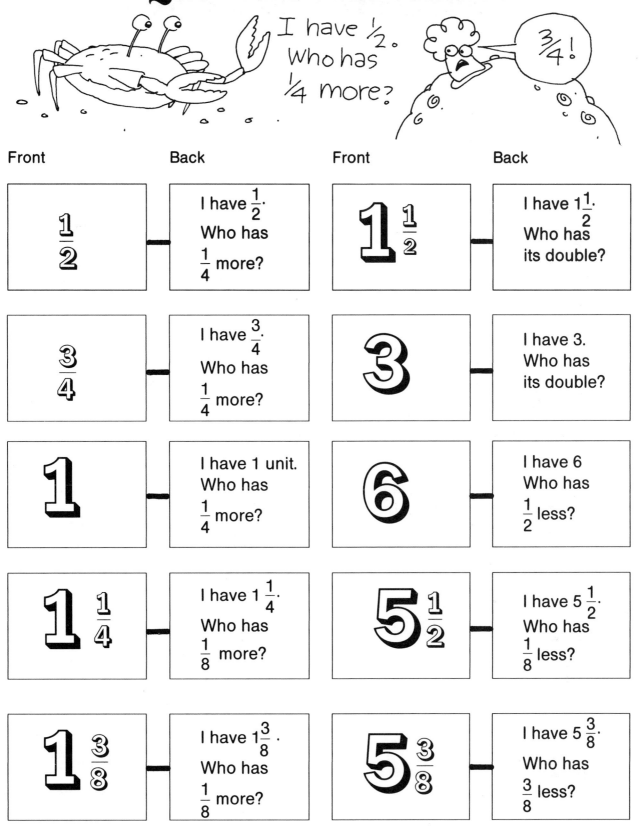

Front	Back	Front	Back
$\frac{1}{2}$	I have $\frac{1}{2}$. Who has $\frac{1}{4}$ more?	$1\frac{1}{2}$	I have $1\frac{1}{2}$. Who has its double?
$\frac{3}{4}$	I have $\frac{3}{4}$. Who has $\frac{1}{4}$ more?	3	I have 3. Who has its double?
1	I have 1 unit. Who has $\frac{1}{4}$ more?	6	I have 6 Who has $\frac{1}{2}$ less?
$1\frac{1}{4}$	I have $1\frac{1}{4}$. Who has $\frac{1}{8}$ more?	$5\frac{1}{2}$	I have $5\frac{1}{2}$. Who has $\frac{1}{8}$ less?
$1\frac{3}{8}$	I have $1\frac{3}{8}$. Who has $\frac{1}{8}$ more?	$5\frac{3}{8}$	I have $5\frac{3}{8}$. Who has $\frac{3}{8}$ less?

215

GA1392

Fraction Chain: Addition, Subtraction of Units, Eighths, Quarters and Halves

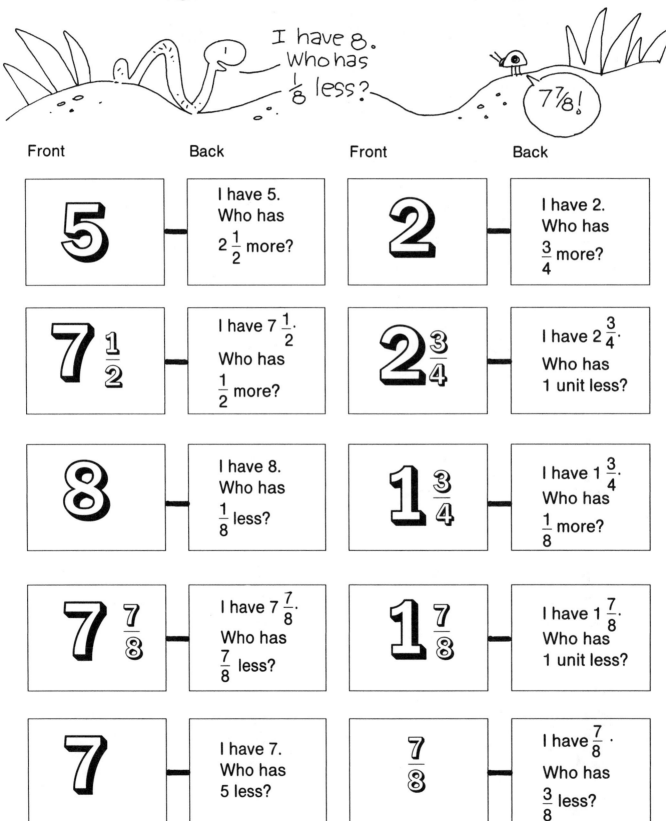

I have 8.
Who has $\frac{1}{8}$ less?

$7\frac{7}{8}$!

Front	Back	Front	Back
5	I have 5. Who has $2\frac{1}{2}$ more?	**2**	I have 2. Who has $\frac{3}{4}$ more?
7$\frac{1}{2}$	I have $7\frac{1}{2}$. Who has $\frac{1}{2}$ more?	**2$\frac{3}{4}$**	I have $2\frac{3}{4}$. Who has 1 unit less?
8	I have 8. Who has $\frac{1}{8}$ less?	**1$\frac{3}{4}$**	I have $1\frac{3}{4}$. Who has $\frac{1}{8}$ more?
7$\frac{7}{8}$	I have $7\frac{7}{8}$. Who has $\frac{7}{8}$ less?	**1$\frac{7}{8}$**	I have $1\frac{7}{8}$. Who has 1 unit less?
7	I have 7. Who has 5 less?	**$\frac{7}{8}$**	I have $\frac{7}{8}$. Who has $\frac{3}{8}$ less?

GA1392

Place Value Chain

Tens, ones

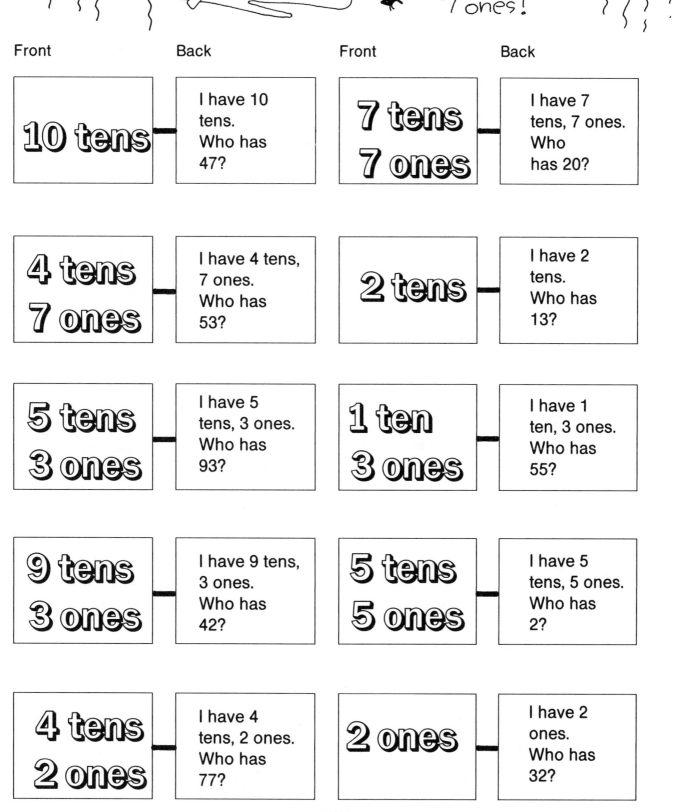

I have 10 tens.
Who has 47?

4 tens,
7 ones!

Front	Back	Front	Back
10 tens	I have 10 tens. Who has 47?	**7 tens 7 ones**	I have 7 tens, 7 ones. Who has 20?
4 tens 7 ones	I have 4 tens, 7 ones. Who has 53?	**2 tens**	I have 2 tens. Who has 13?
5 tens 3 ones	I have 5 tens, 3 ones. Who has 93?	**1 ten 3 ones**	I have 1 ten, 3 ones. Who has 55?
9 tens 3 ones	I have 9 tens, 3 ones. Who has 42?	**5 tens 5 ones**	I have 5 tens, 5 ones. Who has 2?
4 tens 2 ones	I have 4 tens, 2 ones. Who has 77?	**2 ones**	I have 2 ones. Who has 32?

GA1392

Place Value Chain

Tens, ones

4 tens, 6 ones!

I have 4 tens, 6 ones. Who has 29?

Front   Back   Front   Back

| 3 tens 2 ones | I have 3 tens, 2 ones. Who has 46? | 3 tens | I have 3 tens. Who has 90? |

| 4 tens 6 ones | I have 4 tens, 6 ones. Who has 29? | 9 tens | I have 9 tens. Who has 14? |

| 2 tens 9 ones | I have 2 tens, 9 ones. Who has 81? | 1 ten 4 ones | I have 1 ten, 4 ones. Who has 62? |

| 8 tens 1 one | I have 8 tens, 1 one. Who has 76? | 6 tens 2 ones | I have 6 tens, 2 ones. Who has 19? |

| 7 tens 6 ones | I have 7 tens, 6 ones. Who has 30? | 1 ten 9 ones | I have 1 ten, 9 ones. Who has 100? |

218

Multiplication Chain

I have 6. Who has 5×5?   Facts with products to 36

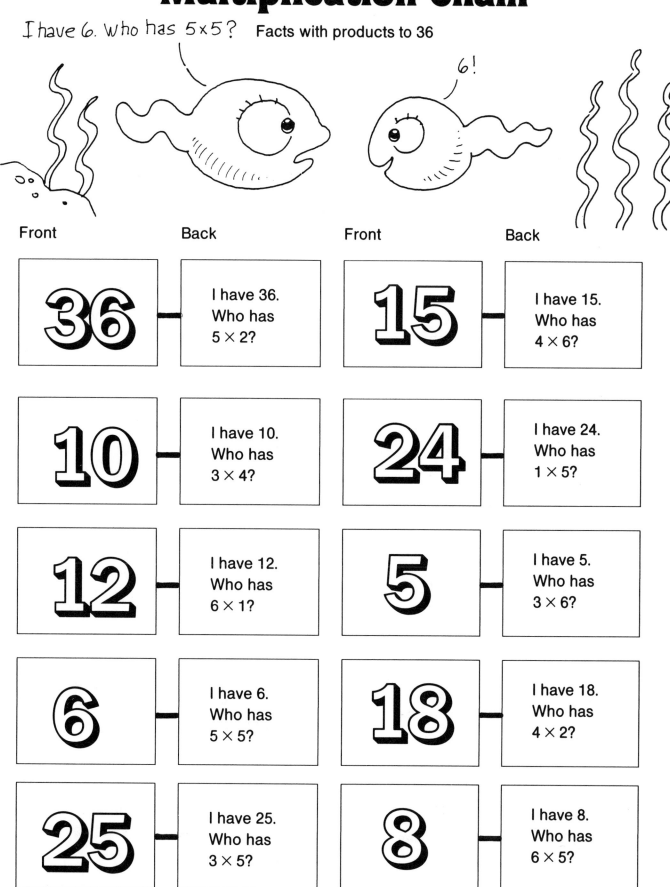

Front	Back	Front	Back
36	I have 36. Who has 5 × 2?	**15**	I have 15. Who has 4 × 6?
10	I have 10. Who has 3 × 4?	**24**	I have 24. Who has 1 × 5?
12	I have 12. Who has 6 × 1?	**5**	I have 5. Who has 3 × 6?
6	I have 6. Who has 5 × 5?	**18**	I have 18. Who has 4 × 2?
25	I have 25. Who has 3 × 5?	**8**	I have 8. Who has 6 × 5?

GA1392

Multiplication Chain

Facts with products to 36

six times zero?

Front	Back	Front	Back
30	I have 30. Who has 4 × 5?	**0**	I have 0. Who has 4 × 1?
20	I have 20. Who has 4 × 4?	**4**	I have 4. Who has 1 × 1?
16	I have 16. Who has 3 × 1?	**1**	I have 1. Who has 2 × 7?
3	I have 3. Who has 3 × 3?	**14**	I have 14. Who has 1 × 2?
9	I have 9. Who has 6 × 0?	**2**	I have 2. Who has 6 × 6?

GA1392